AS/A-LEVEL YEARS 1 AND 2

STUDENT GUIDE

AQA

Chemistry

Practical assessment

Nora Henry

Series editor: Dave Scott

HODDER
EDUCATION
AN HACHETTE UK COMPANY

Hodder Education, an Hachette UK company, Blenheim Court, George Street, Banbury,
Oxfordshire OX16 5BH

Orders

Bookpoint Ltd, 130 Park Drive, Milton Park, Abingdon, Oxfordshire OX14 4SE

tel: 01235 827827

fax: 01235 400401

e-mail: education@bookpoint.co.uk

Lines are open 9.00 a.m.–5.00 p.m., Monday to Saturday, with a 24-hour message answering
service. You can also order through the Hodder Education website: www.hoddereducation.co.uk

© Nora Henry 2017

ISBN 978-1-4718-8514-3

First printed 2017

Impression number 7 6 5

Year 2021 2020 2019 2018

This guide has been written specifically to support students preparing for the AQA AS and
A-level Chemistry examinations. The content has been neither approved nor endorsed by AQA
and remains the sole responsibility of the author.

Cover photo: Ryan McVay/Photodisc/Getty Images/Professional Science 72

Typeset by Integra Software Services Pvt. Ltd, Pondicherry, India

Printed in Dubai

Hachette UK's policy is to use papers that are natural, renewable and recyclable products and
made from wood grown in sustainable forests. The logging and manufacturing processes are
expected to conform to the environmental regulations of the country of origin.

Contents

Skills Guidance

Questions & Answers

■About this book

This guide is written for the AQA AS and A-level Chemistry specifications. Its aim is to help you understand the different practical skills that you are expected to develop throughout the course. These skills are assessed throughout the written examinations and also through the practical endorsement.

Chemistry is a practical subject and practical skills are fundamental to understanding the nature of chemistry. Chemistry gives learners many opportunities to develop the fundamental skills needed to collect and analyse empirical data. In the written papers the following written practical skills are assessed:

- **Independent thinking** — applying scientific knowledge to practical contexts, and solving problems set in practical contexts.
- **Application of scientific methods and practices** — identification of variables, including those that must be controlled, commenting on experimental design and evaluating scientific methods and results. Presenting data in appropriate ways and drawing conclusions with reference to measurement uncertainty is also important.
- **Numeracy and the application of mathematical concepts in a practical context** — processing and analysing data, considering margins of error, accuracy and precision, and plotting and interpreting graphs.
- **Using instruments and equipment** — knowing and understanding how to use a wide range of experimental and practical instruments, equipment and techniques.

Questions are included throughout this book to help you develop these skills. 15% of your examination will be questions on practical skills.

The guide is divided into two sections:

- **Skills Guidance** begins with a look at health and safety issues, followed by a brief guide to using significant figures. (Other mathematical skills are identified in the rest of this section.) This is followed by detailed guidance on each of the 12 required practical activities, which reflects the AQA requirement that you should acquire competence and confidence in a variety of practical, mathematical and problem-solving skills, and in handling apparatus competently and safely. You should be able to design and carry out both the core practical activities and your own investigations. All the required practical activities in this section include worked examples to allow you to think more deeply about why different steps are carried out. Working conscientiously through these examples will give you practice at answering the sort of questions you will be asked in your examinations. Note: for the AS examination you need only be familiar with required practicals 1–6.
- The **Questions & Answers** section pulls together the Skills Guidance content through a range of exam-style questions aimed at testing your practical knowledge. Answers are provided as well as comments to help you to understand the way to approach each question.

The purpose of this guide is to help you understand different practical skills in chemistry, and enable you to answer A-level papers, but don't forget that what you are doing is learning chemistry. The development of an understanding of chemistry can only evolve with experience, which means time spent thinking about chemistry, applying it to unfamiliar situations and solving problems. This guide provides you with a platform for doing this.

If you are reading this, you are clearly determined to do well in your examinations. If you try all the knowledge check questions, worked examples and the questions in the Questions & Answers section before looking at the answers, you will begin to develop understanding and the necessary techniques for answering practical examination questions. As the answers to the worked examples are an integral part of the learning process, they appear immediately after the questions. You are recommended to cover up the answers before attempting the question — this will improve your ability to answer similar questions in future. If you 'cheat' by looking at the answers first, you are only cheating yourself!

Thus prepared, you will be able to approach the examination with confidence.

Skills Guidance

▮ Health and safety

Whenever experiments and investigations are carried out in the laboratory, there are general safety rules that you must follow:

- Wear eye protection at all times.
- Take care when handling hot apparatus.
- Ensure there are no naked flames in the laboratory when using flammable substances.
- Dispose of chemicals as directed — take particular care in disposing of organic chemicals.

When planning an investigation you need to decide if the experiment is safe, by carrying out a **risk assessment**.

A good risk assessment includes:

- a list of all the hazards and why they are hazardous
- a list of the potential risks relating to what you do in the experiment
- suitable control measures you could take that will reduce or prevent the risks

You must recognise the Control of Substances Hazardous to Health (COSHH) hazard warning signs shown in Figure 1.

A **risk assessment** is a judgment of how likely it is that someone might come to harm if a planned action is carried out.

Dangerous to the environment	Toxic	Gas under pressure
Corrosive	Explosive	Flammable

Figure 1 Hazard warning signs →

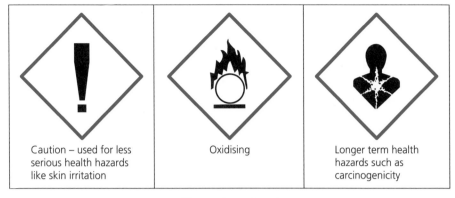

| Caution – used for less serious health hazards like skin irritation | Oxidising | Longer term health hazards such as carcinogenicity |

Figure 1 *continued*

In examinations you may encounter questions about safety applied to particular experiments, and examples of these are included throughout this guide.

Hazard and risk

All practical tasks described in this guide should be risk assessed by a qualified teacher before being performed either as a demonstration or as a class practical. Safety goggles and a laboratory coat or apron must be worn where it is appropriate to do so. The author and the publisher cannot accept responsibility for safety.

■ Significant figures

Practical work leads to various types of calculation, many of which require the use of significant figures. You may find it useful to refer back to this section when you are trying questions later in the book.

Significant figures are those numbers that carry meaning and contribute to the number's precision. The first significant figure of a number is the first digit that is not a zero. In chemistry we quote values to a limited number of significant figures, as we cannot be sure of the exact value to a greater number of significant figures.

The rules for significant figures are as follows:

1 Always count non-zero digits. For example 21 has two significant figures and 8.923 has 4.

2 Never count zeros at the start of a number (leading zeros) even when there is a decimal point in the number. For example 021, 0021 and 0.0021 all have two significant figures.

3 Always count zeros that fall between two non-zero digits. For example 20.8 has three significant figures and 0.001 030 04 has six significant figures.

4 When a number with no decimal point ends in several zeros, these zeros may or may not be significant. The number of significant figures should then be stated. For example: 20 000 (to 3 s.f.) means that the number has been measured to the nearest 100 while 20 000 (to 4 s.f.) means that the number has been measured to the nearest 10.

Knowledge check 1

In an experiment a reaction occurs between dilute sodium hydroxide solution and dilute hydrochloric acid solution. When the reaction is finished, is it safe to dispose of the mixture down the sink?

The rules are best illustrated with some examples:

- 34.23 has four significant figures — always count non-zero digits.
- 6000 has no decimal point and ends in several zeros so it is difficult to say if the zeros are significant. With zeros at the end of a number, the number of significant figures should be stated.
- 2000.0 has a decimal point, hence it has five significant figures.
- 0.036 has two significant figures — never count the zeros at the start of a number even when there is a decimal point.
- 3.0212 has five significant figures.

In calculations you should round the answer to a certain number of significant figures.

The rules for rounding are as follows:

- If the next number is 5 *or more*, round up.
- If the next number is *4 or less*, do not round up.

Worked example 1

Calculate the value of 7.799 g – 6.250 g and give your answer to three significant figures.

Answer

Your calculation would yield 1.549 g. If the answer is required to three significant figures, this would be 1.55 g, because the digit 9 is greater than 5.

When combining measurements with different degrees of accuracy and precision, *the accuracy of the final answer can be no greater than the least accurate measurement*. This means that when measurements are multiplied or divided, the answer can contain no more significant figures than the least accurate measurement.

Worked example 2

In a titration 20.5 cm^3 of 0.25 mol dm^{-3} sodium hydroxide solution reacts with 1.2 mol dm^{-3} hydrochloric acid. When calculating the volume of hydrochloric acid required to neutralise the sodium hydroxide solution, how many significant places should you give your answer to?

Answer

It is useful to look at all the data and write down the number of significant figures in each measurement.

Measurement	Number of significant figures
20.5 cm^3	Three
0.25 mol dm^{-3}	Two
1.2 mol dm^{-3}	Two

The least accurate measurement is to two significant figures, hence when you calculate your answer you must give it to two significant figures.

Knowledge check 2

Use
$$\text{density} = \frac{\text{mass}}{\text{volume}}$$
to calculate the density of a block of iron that has mass of 40.52 g and volume of 5.1 cm^3. Give your answer to an appropriate number of significant figures.

Tip

In calculations such as this one, which have intermediate answers, it is best to leave these in the calculator and use these in subsequent steps, rather than rounding answers at each step. This will minimise rounding errors in the final answer.

Significant figures and standard form

For numbers in scientific form, to find the number of significant figures ignore the exponent (n number) and apply the usual rules.

For example, 6.2091×10^{28} has five significant figures and 1.3×10^2 has two significant figures.

The same number of significant figures must be kept when converting between ordinary and standard form. For example:

$0.0050\,\text{mol}\,\text{dm}^{-3} = 5.0 \times 10^{-3}\,\text{mol}\,\text{dm}^{-3}$ (2 s.f.)

$40.06\,\text{g} = 4.006 \times 10^1\,\text{g}$ (4 s.f.)

$90.0\,\text{g} = 9.00 \times 10^1\,\text{g}$ (3 s.f.)

$0.01070\,\text{kg} = 1.070 \times 10^{-2}\,\text{kg}$ (4 s.f.)

The number 260.99 rounded to:

- 4 s.f. is 261.0
- 3 s.f. is 261
- 2 s.f. is 260
- 1 s.f. is 300

Using standard form makes it easier to identify significant figures. In the example above, 261 has been rounded to the two-significant-figure value of 260. However, if seen in isolation, it would be impossible to know whether the final zero in 260 is significant (and the value to three significant figures) or insignificant (and the value to two significant figures). Standard form, however, is unambiguous:

- 2.6×10^2 is to two significant figures.
- 2.60×10^2 is to three significant figures.

Worked example 3

Convert 0.002 350 to standard form, ensuring that you retain the correct number of significant figures.

Answer

Step 1: Note how many significant figures are present in 0.002 350. Remember that the zeros after the point are not significant so there are four significant figures.

Step 2: To write the number in standard form, write the four significant figures, with a decimal place after the first number, and then write '$\times 10^n$' after it:

2.350×10^n

Step 3: Count how many places the decimal point has moved to the right and write this value as the n value. The n is negative because the decimal point has moved to the right instead of the left:

$0.002350 = 2.350 \times 10^{-3}$

> **Tip**
>
> Many calculations will ask you to give your answer to a certain number of significant figures *and* in standard form.

■ Required practical activities

Required practical 1

Make up a volumetric solution and carry out a simple acid–base titration

Making up a standard solution

To prepare a **standard solution**, an accurately known mass must be dissolved in deionised water and then made up to an accurate volume using a volumetric flask.

When preparing a standard solution a chemical should be used that:

- does not absorb moisture from, or lose moisture to, the environment
- has an accurately known relative formula mass so that the number of moles dissolved can be determined — a hydrated salt is not used for a standard solution
- is very pure
- has a relatively high relative formula mass so that weighing errors are minimised

The procedure for preparing a standard solution (Figure 2) is as follows:

1 Weigh out an accurate mass of a solid in a clean, dry beaker.
2 Add enough deionised water to dissolve the solid, stirring with a glass rod.
3 Transfer the solution with rinsing to a $250\,cm^3$ volumetric flask using a funnel.
4 Rinse the beaker and glass rod with deionised water and add to the volumetric flask.
5 Make up to the mark by adding deionised water until the bottom of the meniscus is on the mark.
6 Stopper the flask and invert to mix thoroughly.

A **standard solution** is a solution for which the concentration is accurately known.

Knowledge check 3

Why is sodium hydroxide not a suitable solid for making up a standard solution?

Tip

When making up a standard solution, a larger mass of solid dissolved gives a smaller weighing error, so it will be a more accurate solution.

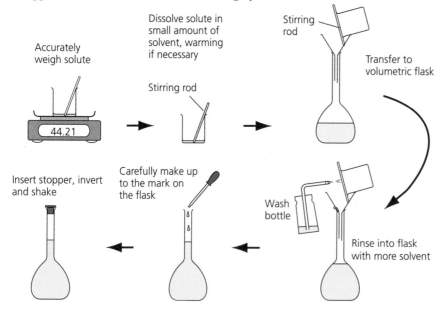

Figure 2 Making up a standard solution

Uncertainty in mass measurements

Uncertainty is an estimate attached to a measurement that gives the range of values within which the true value is thought to lie. This range is normally expressed around a central value, for example 52.0 ± 0.2.

If a balance measures to one decimal place, the mass will normally have an uncertainty of $\pm 0.1\,g$. If a balance measures to two decimal places, the mass will normally have an uncertainty of $\pm 0.01\,g$.

Percentage error is calculated using:

$$\text{percentage error} = \frac{\text{uncertainty}}{\text{quantity measured}} \times 100$$

Table 1 compares the percentage error in weighing 1 g using both balances.

Table 1

Balance to one decimal place	Balance to two decimal places
% error = $\frac{0.1}{1} \times 100 = 10\%$	% error = $\frac{0.01}{1} \times 100 = 1\%$

The *more* decimal places a balance reads to, the *smaller* the percentage error.

When weighing out a *smaller* mass, the percentage error is *more significant*. Weighing out 0.1 g on a balance to two decimal places gives a percentage error of $(0.01/0.1) \times 100 = 10\%$ compared with 1% when weighing out the larger mass of 1 g.

Note that if two mass measurements are taken, as in many experiments, then the formula for the percentage uncertainty is:

$$\text{percentage uncertainty} = \frac{2 \times \text{uncertainty in each measurement}}{\text{quantity measured}} \times 100$$

Knowledge check 4

In an experiment to find the mass of water lost on heating a solid, the following results were obtained:

mass of crucible + solid before heat = 25.45 g; uncertainty = 0.005 g

mass of crucible + solid after heat = 24.21 g; uncertainty = 0.005 g

Calculate the mass of water lost and the percentage uncertainty in this value.

> **Tip**
>
> To reduce the percentage error in measuring the mass of water removed in this experiment record the mass to two decimal places or use a larger mass of hydrated crystals.

Titration method

The titration procedure (Figure 3) is summarised as follows:

1 Rinse the burette with the solution you are going to fill it with. Discard the rinsings and fill the burette.

2 Rinse a pipette with the solution you are going to pipette into the conical flask. Using a pipette and pipette filler transfer $25.0\,cm^3$ of this solution into a conical flask.

3 Add 2–3 drops of a suitable indicator to the conical flask.

4 Add the solution from the burette, with constant swirling, until the indicator just changes colour. This is a trial titration.

> **Tip**
>
> Measuring cylinders have limited accuracy and are not appropriate for use in titrations when volumes to one decimal place are needed.

5   To reduce the effect of random error on titration results, repeat the titration to achieve 2–3 **concordant** results, adding the solution dropwise near the end point.

6   Calculate the mean titre from the concordant results.

Concordant titre results are those that agree to within 0.1 cm³.

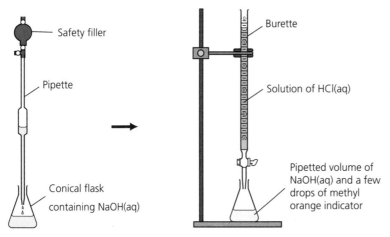

Figure 3 The apparatus used for a titration

> **Tip**
>
> Make sure you remove the funnel used to fill the burette before titrating.

> **Tip**
>
> When carrying out a titration it is useful to keep the solution from the first accurate titration to achieve consistency by colour matching.

Titration results should be recorded in a table. To calculate the mean titre, choose those titres that are concordant — i.e. agree to within 0.1 cm³. Where this is not possible, the two titres that have the closest agreement should be used. Readings that are not concordant are often called outliers, for example the value 22.85 in Table 2.

Table 2

	Trial	1	2	3
Final burette reading/cm³	22.90	45.50	22.85	45.55
Initial burette reading/cm³	0.00	22.90	0.00	22.85
Titre/cm³	22.90	22.60	22.85	22.70
Titres used to calculate mean		✓		✓
Mean titre	22.65 cm³			

> **Tip**
>
> Titration results should be recorded to two decimal places with the last figure either 0 or 5.

When ungraduated glassware, such as a volumetric flask or a pipette, is manufactured, the maximum uncertainty is usually marked on the glassware, as shown in Figure 4.

The uncertainty can be determined for each of the measuring instruments used in a titration — pipette, volumetric flask and burette.

$$\% \text{ uncertainty} = \frac{\text{instrument uncertainty}}{\text{quantity measured}} \times 100$$

Worked example 1

The uncertainty associated with a pipette reading is ±0.06 cm³. Calculate the percentage uncertainty when using a 25.0 cm³ pipette.

Answer

$$\% \text{ uncertainty} = \frac{\text{instrument uncertainty}}{\text{quantity measured}} \times 100 = (0.06/25.0) \times 100 = 0.24\%$$

Figure 4 A 15 ml pipette (1 ml = 1 cm³); the uncertainty is marked as ±0.03

Worked example 2

a A burette has uncertainty $\pm0.05\,cm^3$. In a titration, the initial burette reading was $0.05\,cm^3$ and the final burette reading was $22.55\,cm^3$. What is the percentage uncertainty in the titre value?

b On repeating the experiment, a mean titre value of $22.45\,cm^3$ was recorded. The uncertainty in the mean titre was ±0.15. Calculate the percentage uncertainty in the mean titre.

Answer

a titre value = final burette reading – initial burette reading = 22.55 – 0.05
= $22.50\,cm^3$

The overall uncertainty in any volume measured in a burette always comes from the two measurements, so to find the overall uncertainty multiply the uncertainty by two.

$$\text{percentage uncertainty} = \frac{2 \times 0.05}{22.50} \times 100 = 0.44\%$$

b There is only one value given, the mean titre.

$$\text{percentage uncertainty} = \frac{0.15}{22.45} \times 100 = 0.67\%$$

Tip

When calculating the uncertainty for a titre reading remember that the uncertainty applies for each reading, so 2 × instrument uncertainty is used.

Worked example 3

In an experiment a student made two different measurements: $25\,cm^3$ of hydrochloric acid using a measuring cylinder with uncertainty $\pm1\,cm^3$ and $0.40\,g$ of calcium carbonate using a balance with uncertainty $\pm0.01\,g$. State and explain whether the balance or the measuring cylinder contributed most to the measurement errors in this experiment.

Answer

$$\text{percentage uncertainty of cylinder} = \frac{1}{25} \times 100 = 4\%$$

$$\text{percentage uncertainty of balance} = \frac{0.01}{0.40} \times 100 = 2.5\%$$

The measuring cylinder contributed most to the measurement errors.

Knowledge check 5

What is the colour change of phenolphthalein indicator when a solution of sodium hydroxide solution of unknown concentration is titrated with standard sulfuric acid solution?

The choice of indicator used in a titration depends on the type of titration (Table 3).

Table 3

Indicator	Colour in acid	Colour in alkali	Titrations suitable for
Methyl orange	Red	Yellow	Strong acid–strong base Strong acid–weak base
Phenolphthalein	Colourless	Pink	Strong acid–strong base Weak acid–strong base

Some of the solution from the burette may run down the inside of the conical flask, rather than into the solution it contains. The addition has been measured by the burette, so you must make sure that all the solution added has a chance to react. Sometimes, you can achieve this by swirling the solutions together but deionised water

Tip

Remember that clear is not a colour. Phenolphthalein is colourless in acid; it is also clear, but colourless describes the colour.

can also be squirted down the side of the flask to wash everything into the solution below. Although the solution is undoubtedly diluted, the amount, in moles, present in the pipetted solution will not have been changed and it is this that is being titrated.

Worked example 4

In a titration $25.0\,cm^3$ of potassium hydroxide solution is added to a conical flask using a pipette. Three drops of methyl orange indicator are added and nitric acid added from a burette until the indicator changes colour. Which of the following would lead to the titre being larger than it should be?

A Rinsing the conical flask with water before adding the potassium hydroxide solution.

B Rinsing the burette with water before filling it with hydrochloric acid.

C Rinsing the pipette with water before filling it with potassium hydroxide solution.

D Adding extra drops of indicator.

Answer

- Rinsing the burette with water before use will make the hydrochloric acid in the burette more dilute, and so more of it will be needed for reaction, and so a larger titre value.
- Rinsing the pipette/conical flask with water would dilute the potassium hydroxide, so less hydrochloric acid would need to be added for neutralisation and a smaller titre value.
- Adding extra drops of indicator would not have an effect.

Hence the answer is B.

Worked example 5

Describe how you would dilute $25.0\,cm^3$ of oven cleaner to $250\,cm^3$ and safely transfer $25.0\,cm^3$ of the diluted solution to a conical flask.

Answer

- Rinse a pipette with oven cleaner.
- Use a safety pipette filler with the pipette to draw up $25.0\,cm^3$ of oven cleaner and place into a $250\,cm^3$ volumetric flask.
- Make up to the mark by adding deionised water until the bottom of the meniscus is on the mark.
- Stopper the flask and invert to mix thoroughly.
- Rinse a pipette with the diluted solution.
- Transfer $25.0\,cm^3$ of diluted solution using the pipette and safety filler to a conical flask.

Sometimes solutions that are made up in volumetric flasks are not standard, for example an indigestion tablet can be reacted with a known excess of acid and the solution made up to $250\,cm^3$ in a volumetric flask. The method is similar — the tablet is weighed in the beaker, a volume of acid added and the solution added to a volumetric flask, with rinsings, and made up to $250\,cm^3$. Portions of this solution can then be titrated to determine, for example, the mass of calcium carbonate in the indigestion tablet. Worked examples 6 and 7 outline this type of titration, which is often referred to as a **back titration**.

Worked example 6

A back titration was used to calculate the mass of calcium carbonate in an indigestion tablet. Solution A was made by reacting *two* indigestion tablets, total mass $2.64\,g$, with $25.0\,cm^3$ of $2.00\,mol\,dm^{-3}$ hydrochloric acid (an excess) and then making this up to $250\,cm^3$.

$25.0\,cm^3$ of this solution was then titrated against $0.100\,mol\,dm^{-3}$ sodium hydroxide solution and the titre was found to be $23.80\,cm^3$.

a The indicator used was phenolphthalein. State the colour change at the end point of the titration.

b Write an equation for the two reactions that occur in this experiment.

c Name the piece of apparatus used to make up and contain the $250\,cm^3$ of solution A.

d i Calculate the mass of calcium carbonate present in one indigestion tablet.

  ii Calculate the percentage of calcium carbonate present in a $0.850\,g$ tablet.

e i The indigestion tablets also contain glucose, sucrose and a flavouring. Suggest why these do not affect the titration value obtained.

  ii Suggest one way in which the accuracy of the titration can be increased.

  iii Suggest how the reliability can be increased.

Answer

a The conical flask contains the unreacted acid, so the colour change is colourless to pink.

b The calcium carbonate in the indigestion tablets react with hydrochloric acid:

$$CaCO_3 + 2HCl \rightarrow CaCl_2 + H_2O + CO_2$$

The excess hydrochloric acid then reacts with sodium hydroxide in the titration:

$$HCl + NaOH \rightarrow NaCl + H_2O$$

c $250\,cm^3$ volumetric flask

d i moles of sodium hydroxide $= \dfrac{23.80 \times 0.100}{1000} = 0.00238$

The ratio is $1 : 1$, so:

moles of hydrochloric acid in $25.0\,cm^3$ of solution A $= 0.00238$

moles of unreacted hydrochloric acid in $250\,cm^3$ solution A
$= 0.00238 \times 10 = 0.0238$

moles of hydrochloric acid added to the indigestion tablets
$= \dfrac{25.0 \times 2.00}{1000} = 0.05$

number of moles of acid that reacted with the tablets
$= 0.05 - 0.0238 = 0.0262$

Tip

When calculating a concentration from a titration it is best to keep intermediate answers in your calculator, and use these in subsequent steps. This will minimise rounding errors in the final answer.

The ratio is $2HCl : 1CaCO_3$

number of moles of calcium carbonate in the indigestion tablets

$$= \frac{0.0262}{2} = 0.0131$$

mass of calcium carbonate in each indigestion tablet

$$= \frac{\text{moles} \times M_r}{2} = \frac{0.0131 \times 100.1}{2} = 0.656\,g$$

ii $\% = \dfrac{\text{mass of calcium carbonate}}{\text{mass of tablet}} \times 100 = \dfrac{0.656}{0.850} \times 100 = 77.2\%$

e i They do not react with the hydrochloric acid.

ii Adding the acid in drops at the end point or washing down the sides of the flask with deionised water during titration.

iii By carrying out further titrations and calculating an average of concordant results. Reliability can also be increased by repeating the complete experiment.

Worked example 7

The relative formula mass of pure $Ca(OH)_2$ can be determined experimentally by reacting a measured mass of the pure solid with an excess of hydrochloric acid. The equation for this reaction is:

$$Ca(OH)_2 + 2HCl \rightarrow CaCl_2 + 2H_2O$$

The unreacted acid can then be determined by titration with a standard sodium hydroxide solution. You are provided with $50.0\,cm^3$ of $0.2\,mol\,dm^{-3}$ hydrochloric acid and $0.300\,g$ calcium carbonate. Show by calculation that this is an appropriate mass to use and outline, giving brief practical details, how you would carry out an experiment to calculate the relative formula mass of the solid.

Answer

number of moles of HCl $= \dfrac{50.0 \times 0.200}{1000} = 0.010$

ratio $= 1Ca(OH)_2 : 2HCl$

amount in moles of $Ca(OH)_2$ needed to react with all of the acid is

$$\frac{0.010}{2} = 0.005$$

amount in moles of calcium hydroxide provided $= \dfrac{\text{mass}}{M_r} = \dfrac{0.300}{74.1}$
$= 0.004\,moles$

This is appropriate because the acid will be in excess.

Method: measure out $0.300\,g$ of calcium carbonate, place in a conical flask and add $50.0\,cm^3$ of hydrochloric acid using a pipette. Add three drops of indicator and titrate against $0.100\,mol\,dm^{-3}$ sodium hydroxide from a burette, until the indicator colour changes.

$\rightarrow$

Calculation: calculate the moles of sodium hydroxide used; by ratio this is the amount in moles of hydrochloric acid left. Divide this by two, because of the ratio, to find the number of moles of calcium hydroxide. Subtract this from the amount in moles of hydrochloric acid initially added. Then calculate M_r using M_r = mass of calcium hydroxide (0.300)/moles of calcium hydroxide.

Required practical 2

Measurement of an enthalpy change

Determination of enthalpy change of combustion

The simple apparatus used to measure the **enthalpy change of combustion** of a liquid fuel is shown in Figure 5. The calorimeter should be insulated from its surroundings and usually contains water.

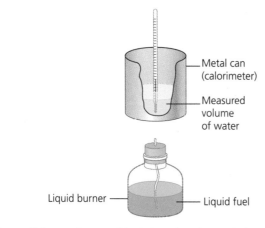

Figure 5 Apparatus used to determine the enthalpy change of combustion of a volatile liquid such as ethanol

An accurate thermometer measures temperature rise and the energy change is calculated using:

heat energy change = mass × specific heat capacity × temperature change

$$q = mc\Delta T$$

The energy change is scaled to find the energy change for 1 mole of fuel.

The method to determine the enthalpy of combustion of a fuel such as ethanol or methanol is as follows:

1 Accurately measure 100 cm³ of water into a calorimeter/beaker. (100 cm³ is the same as 100 g water because the density of water is 1 g cm⁻³.)
2 Weigh a spirit burner containing the liquid to be burnt.
3 Measure the initial temperature of the water using a thermometer (T_1).
4 Use the spirit burner to heat the water.

Enthalpy change of combustion is the enthalpy change when 1 mole of a substance is burned completely in excess oxygen with all reactants and products in their standard states under standard conditions.

Knowledge check 6

Write a balanced symbol equation for the enthalpy of combustion of methane (CH_4).

Tip

The specific heat capacity of water is 4.18 J g⁻¹ K⁻¹. Temperature change may be given in kelvin or in °C.

Knowledge check 7

State three masses that must be recorded during an experiment to determine the enthalpy of combustion of methanol.

5   Stop heating when there is a reasonable temperature rise (15°C). Stir and measure the final temperature (T_2) of the water using a thermometer.

6   Reweigh the spirit burner.

7   Calculate the temperature change using $\Delta T = T_2 - T_1$ and the heat energy change in joules using $q = mc\Delta T$.

8   Calculate the mass of fuel used in the burner by subtraction, and calculate the number of moles of fuel used using moles = mass/M_r.

9   Calculate the energy change per mole of fuel used.

The value for the enthalpy of combustion determined experimentally is frequently *less* exothermic than the value found in data books. Reasons for this are:

■ heat losses to the surroundings from the spirit burner, wick and calorimeter (the flame is affected by draughts)
■ loss of fuel from the wick or burner, by evaporation
■ loss of water by evaporation
■ incomplete combustion of the fuel, leaving soot on the bottom of the calorimeter
■ heat used to raise the temperature of calorimeter
■ the reaction is unlikely to occur under standard conditions, especially temperature

Simple experiments can be *improved* to obtain more accurate values by:

■ using a draught shield to reduce heat loss to the surroundings
■ using a lid on the calorimeter to reduce heat loss to the surroundings
■ minimising the distance between the flame and the calorimeter
■ insulating the calorimeter and the spirit burner to reduce heat loss
■ using a top on the spirit burner, with wick protruding, to minimise evaporation
■ if possible, burning in a supply of pure oxygen to prevent incomplete combustion

Determination of enthalpy change of neutralisation

Enthalpy changes in solution, such as **enthalpy change of neutralisation**, can be measured using insulated plastic cups as calorimeters, as shown in Figure 6.

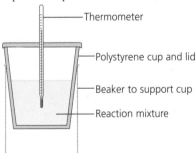

Figure 6 Apparatus used to determine enthalpy change of neutralisation

The method to determine the enthalpy change of neutralisation is as follows:

1   Place a polystyrene cup in a glass beaker for support.

2   Rinse a measuring cylinder with $1.0\,mol\,dm^{-3}$ HCl and then measure $25\,cm^3$ of this acid and transfer into the polystyrene cup.

3   Stir the acid with a thermometer and record the temperature.

Tip

Remember that in the calculation using $q = mc\Delta T$ it is always the mass of water in the calorimeter that is used. Often a volume of water is given, but since the density of water is $1\,g\,cm^{-3}$ and density = mass/volume, then the values of volume and mass are the same.

Knowledge check 8

Why is a draught shield placed around the spirit burner and calorimeter?

Knowledge check 9

In the incomplete combustion of an alcohol, water is produced. Name two other products of this incomplete combustion.

The **enthalpy change of neutralisation** is the enthalpy change when an acid and alkali react to produce 1 mole of water.

4 Rinse a second measuring cylinder with $1.0\,mol\,dm^{-3}$ NaOH(aq) and then measure out $25\,cm^3$ of NaOH(aq).

5 Add the NaOH(aq) to the acid, stir and record the highest temperature reached.

6 Calculate temperature change using $\Delta T = T_2 - T_1$ and the heat energy change in joules using $q = mc\Delta T$.

7 Calculate the amount in mol of acid used, the amount in mol of water formed and the enthalpy of neutralisation.

This method involves measuring the temperature before mixing and the maximum temperature after mixing the reactants, and finding the difference.

An alternative method is to record the temperature against time, and plot a graph, extrapolating the cooling curve after the reaction takes place in order to calculate the temperature change (ΔT) at the point of mixing the reactants, as shown in Figure 7. The extrapolation method is more important when the reaction is highly exothermic because more heat energy is lost at the point of reaction compared with that of a less exothermic reaction. Hence simply measuring the temperature difference may significantly underestimate the temperature change and the value of ΔH.

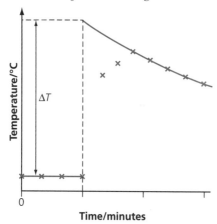

Figure 7 Estimating the maximum temperature of a neutralisation reaction

Worked example

a In an experiment to determine the enthalpy change of combustion of butan-1-ol (C_4H_9OH) the following results were obtained:

mass of burner + butan-1-ol = 55.40 g

mass of burner + butan-1-ol after burning = 53.89 g

initial temperature of water = 20.2°C

final temperature of water = 44.4°C

mass of water in calorimeter = 100 g

Calculate the enthalpy change of combustion.

b In this experiment the student uses a thermometer with uncertainty ±0.1°C in each reading. Calculate the percentage uncertainty in the temperature rise.

Answer

a mass of butan-1-ol burned = $55.40 - 53.89 = 1.510\,g$ M_r butan-1-ol = 74.0

moles of butan-1-ol $= \dfrac{\text{mass}}{M_r} = \dfrac{1.510}{74.0} = 0.0204\,\text{mol}$

energy absorbed by water $= q = mc\Delta T = 100 \times 4.18 \times 24.2 = 10\,116\,J$

energy change of combustion $= \dfrac{10\,116}{0.0204} = -496\,kJ\,mol^{-1}$

b % uncertainty $= \dfrac{\text{instrument uncertainty}}{\text{quantity measured}} \times 100$

$\dfrac{2 \times 0.1}{24.2} \times 100 = 0.8\%$ (1 s.f.)

> **Tip**
>
> In this calculation the answer **a** is given to three significant figures, as this is the number of significant figures of the least accurate measurement (temperature, mass of water).

Required practical 3

Investigation of how the rate of a reaction changes with temperature

> **Tip**
>
> When two readings are taken the uncertainty must be multiplied by 2.

The rate of a reaction can be determined by measuring:

- the loss of a reactant over time
- the gain of a product over time

Investigating the rate is carried out practically in different ways, depending on the reaction being studied. To experimentally investigate reactions that produce a gas the volume of gas produced or the mass of gas lost could be measured over a time period and a graph of volume or mass against time plotted. These methods are looked at in more detail on pages 37–39 and can be used to investigate the rate of a reaction between acid and metal or acid and carbonate.

> **Knowledge check 11**
>
> Why does the rate of the reaction of magnesium and acid increase when the concentration of acid is increased?

When two aqueous solutions are mixed sometimes a precipitate forms. One way of investigating the rate of a precipitation reaction is to place the reaction flask containing one reactant solution on top of a piece of paper with a cross drawn on it. The time is recorded from when the second reactant solution is added until the precipitate forms and obscures the cross from the view of an observer looking down into the flask (Figure 8). The reaction of sodium thiosulfate with acid can be investigated using this method:

$$Na_2S_2O_3(aq) + 2HCl(aq) \rightarrow S(s) + SO_2(g) + H_2O(l) + 2NaCl(aq)$$

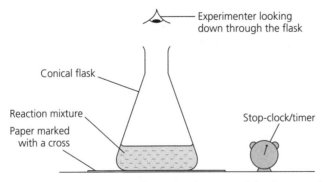

Figure 8 Investigating the rate of reaction for a precipitation reaction

Worked example

25 cm^3 of sodium thiosulfate solution was measured into a conical flask, placed on a cross marked with pencil on paper. 10 cm^3 of hydrochloric acid was added and the stop clock started. The time taken for the sulfur precipitate formed to obscure the cross was recorded.

$$Na_2S_2O_3(aq) + 2HCl(aq) \rightarrow S(s) + SO_2(g) + H_2O(l) + 2NaCl(aq)$$

The experiment was repeated at five different temperatures — the approximate temperatures were room temperature, 25°C, 30°C, 35°C and 40°C. The results obtained are given in Table 4.

Table 4

Average temperature/°C	Time taken for cross to be obscured/s	Rate of reaction/s^{-1}
20	94	
25	67	
30	53	
35	38	
40	29	

a Write an ionic equation for this reaction.

b The quantity of sulfur needed to hide the cross is the same each time, so the rate of the reaction is proportional to 1/time. Calculate the value to two significant figures for the rate of reaction for each temperature.

c The method given was for the reaction at room temperature. Describe how the experiment was carried out at 40°C.

d Plot a graph of rate (y-axis) against temperature (x-axis). Describe the relationship between rate of reaction and temperature shown by your graph. Explain this relationship in terms of collision theory.

e Suggest any safety precautions, apart from wearing safety glasses, that should be followed when carrying out this experiment.

f Suggest any improvements that could be made to the design of this experiment, to improve accuracy.

Answer

a $S_2O_3^{2-}(aq) + 2H^+(aq) \rightarrow S(s) + SO_2(g) + H_2O(l)$

b $1/94 = 0.011 \, s^{-1}$
   $1/67 = 0.015 \, s^{-1}$
   $1/53 = 0.019 \, s^{-1}$
   $1/38 = 0.026 \, s^{-1}$
   $1/29 = 0.034 \, s^{-1}$

c One of the best ways is by placing the flask in a thermostatically controlled water bath at 40°C. Otherwise, heat the sodium thiosulfate solution to 40°C using a Bunsen, tripod and gauze and then place on cross and add acid.

→

> **Tip**
>
> When carrying out this experiment it is not necessary for these exact temperatures to be used, though the temperature must be accurately determined. To do this, measure both the initial temperature and the final temperature and calculate a mean temperature.

> **Tip**
>
> Part d may be more straightforward if you multiply all of the rate values by a common factor (e.g. 10^4).

d

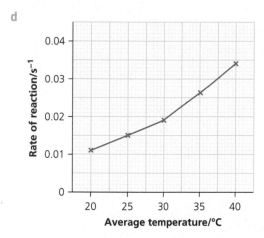

Figure 9 A graph of rate of reaction against temperature

As the temperature increases, the rate of reaction increases. The particles gain energy and move faster, meaning that there are more collisions with the activation energy per unit time.

e Sulfur dioxide is a toxic gas that aggravates asthmatics. Use a fume cupboard. As soon as the reaction is complete pour the solutions away, preferably into the fume cupboard sink. Wash away with plenty of water. This is particularly important with solutions used at higher temperatures.

f A light sensor placed under the flask and linked to a computer could be used, instead of the experimenter's eye.

Required practical 4

Carry out simple test-tube reactions to identify cations and anions

Qualitative testing for ions is usually carried out on a test-tube scale. Many of these tests rely on the formation of a precipitate (ppt). A precipitate is a solid formed when two ionic solutions are mixed. It is formed because one of the combinations of ions from the solutions gives an insoluble compound. For example, when barium chloride solution is added to sodium sulfate solution the barium ions and the sulfate ions react and a precipitate of insoluble barium sulfate forms:

$$Ba^{2+}(aq) \quad + \quad SO_4^{2-}(aq) \quad \rightarrow \quad BaSO_4(s)$$

from barium   from sodium
chloride   sulfate

The methods for carrying out various ion tests are shown in Table 5.

Knowledge check 12

Describe how you would prove that a sample of a fertiliser contained ammonium ions.

Table 5

Ion	Experimental method	Result
Cation tests		
Ammonium (NH$_4^+$)	*Warm* the sample with **sodium hydroxide solution** in a test tube and test the gas produced with *moist* red litmus/universal indicator paper *Or* test the gas produced with a glass rod dipped in concentrated HCl(aq)	Pungent-smelling gas changes moist red litmus paper/universal indicator paper to blue. The gas is alkaline (ammonia) White fumes of ammonium chloride indicate ammonia produced This indicates that the ion in the salt was ammonium
Magnesium (Mg^{2+})	Make a solution of the compound and add a few drops of **sodium hydroxide solution**, and then excess sodium hydroxide solution	White ppt
Calcium (Ca^{2+})	Dip a nichrome wire in concentrated hydrochloric acid and then into the sample. Place in a blue Bunsen flame and record the colour of the flame	Brick-red flame
Barium (Ba^{2+})	Dip a nichrome wire in concentrated hydrochloric acid and then into the sample. Place in a blue Bunsen flame and record the colour of the flame	Green flame
Strontium (Sr^{2+})	Dip a nichrome wire in concentrated hydrochloric acid and then into the sample. Place in a blue Bunsen flame and record the colour of the flame.	Red flame
Anion tests		
Chloride (Cl$^-$)	Make a solution of the compound. Add some **acidified silver nitrate** solution followed by aqueous ammonia	White ppt, which dissolves in dilute aqueous ammonia to give a colourless solution
Bromide (Br$^-$)	Make a solution of the compound. Add some **acidified silver nitrate** solution followed by aqueous ammonia	Cream ppt, which is insoluble in dilute aqueous ammonia but dissolves in concentrated aqueous ammonia to give a colourless solution
Iodide (I$^-$)	Make a solution of the compound. Add some **acidified silver nitrate** solution followed by ammonia solution	Yellow ppt, which is insoluble in dilute and concentrated aqueous ammonia
Sulfate (SO$_4^{2-}$)	Make a solution of the compound. Add some **acidified barium chloride solution**	White ppt
Hydroxide (OH$^-$)	Warm with some solid ammonium salt and test the gas produced with red litmus paper/glass rod dipped in concentrated HCl(aq)	Pungent gas (ammonia) released, which changes red litmus to blue and produces white fumes with concentrated HCl(aq) This is the reverse of the test for ammonium ion
Carbonate (CO$_3^{2-}$)	Add some dilute nitric acid and test the gas produced with limewater	Effervescence, and the gas changes colourless limewater milky

In the test for halide ions the silver nitrate solution is acidified using nitric acid. The silver nitrate removes other ions such as carbonate ions (CO$_3^{2-}$) and hydroxide ions (OH$^-$), which may be present and would react with the silver ions in the silver nitrate solution, forming precipitates that would interfere with the test.

In the test for sulfate ions, the barium chloride solution is often acidified with hydrochloric or nitric acid which reacts with and removes any carbonate ions present. Barium carbonate is a white insoluble solid that would be indistinguishable from barium sulfate.

Knowledge check 13

What is observed when a solution of sulfuric acid is added to some barium nitrate solution?

It is important to note that these tests can be used the other way round. For example, to identify two solutions as nitric acid or silver nitrate a carbonate solution and a chloride solution can be used, giving the results shown in Table 6.

Table 6

Solution	Add sodium carbonate solution	Add sodium chloride solution
Nitric acid	Effervescence	No reaction
Silver nitrate	White ppt	White ppt

It is also useful, in identification, to remember that the solubility of group 2 hydroxides can be investigated by mixing solutions of soluble group 2 salts with sodium hydroxide solution. The solubility *increases* down the group. The solubility of group 2 sulfates can be investigated by mixing solutions of soluble group 2 salts with sulfuric acid. The solubility *decreases* down the group.

Worked example 1

Four solutions to be identified in the laboratory are:

- ammonium sulfate
- potassium sulfate
- potassium iodide
- sodium carbonate

In a practical $1\,cm^3$ of each solution was added to separate test tubes and $1\,cm^3$ of dilute hydrochloric acid added. Effervescence occurred in one test tube and the gas was collected and bubbled into colourless limewater, which changed to milky.

Describe how you would continue this experiment to positively identify all of the solutions.

Answer

The reaction described identifies the sodium carbonate solution.

A good way of planning your work is to quickly draw a table showing the solutions you use and the observations that would occur (Table 7). Of the three remaining substances two are sulfates, which could be identified using barium chloride solution, and then distinguished between by testing for ammonium ion. The iodide ion can be tested for by adding acidified silver nitrate solution.

→

Tip

When choosing a carbonate, chloride or bromide to use as a test solution, choose a group 1 compound, as these are soluble.

Knowledge check 14

What is observed when a few drops of a solution of sodium hydroxide are added to magnesium chloride solution, and then to barium chloride solution?

Table 7

Solution	Add acidified barium chloride solution	Warm with sodium hydroxide solution	Add acidified silver nitrate solution followed by dilute aqueous ammonia
Potassium sulfate	White ppt	No reaction	
Ammonium sulfate	White ppt	Gas released, which changes indicator paper blue	
Potassium iodide	No reaction		Yellow ppt

Description of experiment

$1 \, cm^3$ of the remaining three solutions is placed in separate test tubes and some acidified barium chloride solution added. A white precipitate is formed in two of the test tubes. Fresh $1 \, cm^3$ samples of these two solutions are warmed with dilute sodium hydroxide solution.

In one test tube a gas is produced and is tested with moist universal indicator paper, which turns blue, indicating an alkaline gas produced and that this solution is ammonium sulfate. The other test tube is the potassium sulfate solution.

To the final unidentified sample acidified silver nitrate solution is added, followed by dilute aqueous ammonia; a yellow ppt insoluble in dilute aqueous ammonia is produced, identifying the solution as potassium iodide.

Worked example 2

Another style of question is to complete deductions from given observations. Look carefully at the test column, and note the reagent added. This should aid you in making a deduction. In the example shown (Table 8) the reagents have been highlighted in bold.

Table 8

Test	Observations	Deductions
1 Add a spatula measure of Y to a test tube one third full of **sodium hydroxide solution** and *warm* gently. Carefully smell any gas given off and test it with moist universal indicator paper.	Universal indicator turns blue Pungent/choking smell	Basic gas produced Ammonia gas An ammonium cation could be present
2 Add a spatula measure of Y to a test tube containing $1 \, cm^3$ of dilute **nitric acid**. Add four drops of **barium chloride solution**.	No effervescence White ppt forms	A carbonate anion is not present A sulfate anion is present

Required practical 5

Distillation of a product from a reaction — preparation of an organic liquid

The exact method to prepare an organic liquid depends on the actual liquid being prepared but, in general, an organic liquid can be synthesised and purified using several stages, as shown in Figure 10.

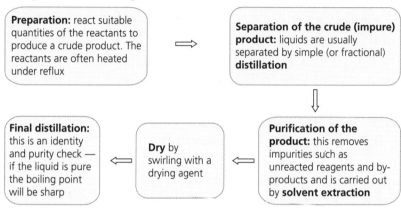

Figure 10 Synthesis of an organic liquid

Preparation

When preparing an organic liquid, the reactions are often slow, as the organic reactants contain strong covalent bonds. It is often necessary to heat the reactants under **reflux** for some time when preparing an organic liquid.

Before reflux the reactants are often added slowly to the reaction flask, with cooling. This is because the reaction is often exothermic and adding reactants slowly with cooling dissipates the heat, preventing the temperature of the reaction increasing and avoiding dangerous splashing or the formation of side-products.

Heating under reflux involves attaching a water-cooled condenser vertically to the reaction flask, as shown in Figure 11. Vapour from the boiling reaction mixture condenses and flows back into the reaction flask. The condenser prevents vapour escaping and so the reactants can be boiled for a long period without any loss of vapour. Note that there is no stopper or thermometer at the top of the apparatus.

Anti-bumping granules are small, rough pieces of silica or unglazed pottery that are added to the reaction mixture before reflux. Most liquids do not boil smoothly — often a large bubble forms at the bottom, and moves abruptly upwards, causing hazardous splashing; this is called bumping. Anti-bumping granules provide a rough surface on which small gas bubbles can grow, hence avoiding bumping. Anti-bumping granules thus promote smooth, even boiling.

Most organic compounds are flammable, so often a flameless method is used to heat them. Instead of a Bunsen burner a water bath, sand bath or electric heating mantle can be used. A water bath is only suitable if the temperature needed is less than 100°C. A electric heating mantle can be used for higher temperatures, and sometimes a sand bath — simply a container with sand in it and heated by a hot

> **Reflux** is the continual boiling and condensing of a reaction mixture to ensure that the reaction takes place without the contents of the flask boiling dry.

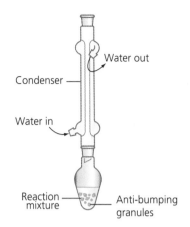

Figure 11 Quickfit apparatus for reflux.
Quickfit has ground-glass joints to ensure a good connection

plate, or a layer of sand in a heating mantle — is used. The sand conducts the heat to the reaction flask and spreads the heat out, so that the flask is heated evenly without the need for stirring.

Some organic compounds are not prepared by refluxing; instead, distillation is used. This is the case for an aldehyde prepared by the oxidation of a primary alcohol by acidified potassium dichromate. The aldehyde is prepared using distillation apparatus, so it is removed from contact with the oxidising agent immediately. If it remained in contact, as it would in reflux, it would be oxidised further to the carboxylic acid.

Separation of the crude product

After reflux, allow the apparatus to cool and rearrange it for **distillation**. To do this the condenser is removed, a still head is inserted in the flask, a thermometer placed in the still head and the water-cooled condenser attached from the side. A receiver is attached to the end of the condenser. Distillation apparatus is shown in Figure 12.

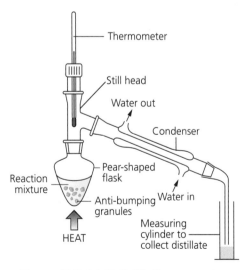

Figure 12 Quickfit distillation apparatus

Anti-bumping granules must also be used for smooth boiling in distillation, because if bumping occurs the liquid can splash over into the condenser, causing an impure product, or can blow the distillation apparatus apart. The distillate is the crude product, and is collected over a boiling point range.

Drawing diagrams

In examinations you may be asked to draw diagrams of the apparatus set up for reflux or for distillation. It is important that you draw these diagrams as cross sections, as shown in Figures 13 and 14, not as 3D pictures as shown in Figures 11 and 12. Diagrams must be labelled. Note that in the distillation apparatus there is a vent on the receiver, so that the apparatus is not closed — an alternative to this is to collect the product in an open measuring cylinder, as shown in Figure 12.

Tip

Diagrams of distillation and reflux do not need to show Bunsen burners, or clamps.

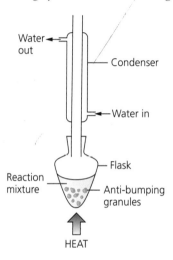

Figure 13 Diagram of reflux

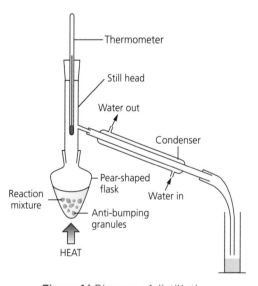

Figure 14 Diagram of distillation

Tip

The bulb of the thermometer is opposite the exit to the condenser so that the temperature of the vapour exiting, which condenses to form the product, can be recorded.

Purification of the product

The crude (impure) liquid product is often contaminated by products or unreacted reactants. Purification of the liquid can be carried out using **solvent extraction** in a separating funnel (Figure 15). This involves shaking the organic liquid with an aqueous solution. The impurities are more soluble in the aqueous solution and move into it — they are extracted. The organic liquid and aqueous solution are immiscible and can be separated.

Immiscible liquids are those that do not mix, and form two layers.

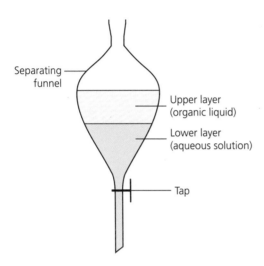

Figure 15 Separating funnel

Solvent extraction

1 Place the organic liquid in a separating funnel and add a portion (e.g. 10 cm³) of aqueous solution (most often this is **sodium hydrogen carbonate solution** to remove acidic impurities in the organic liquid).

2 Stopper and shake, releasing the pressure (often due to build-up of carbon dioxide in an acid + carbonate reaction) by inverting and opening the tap.

3 Allow the separating funnel to stand until the layers settle and separate.

4 Remove the stopper and open the tap, to run off the bottom layer into a beaker. When the bottom layer is nearly all drained close the tap partially to slow down the flow and avoid any of the top layer leaving the funnel. Run off the second layer into a separate beaker.

5 Discard the aqueous layer.

6 Place the organic layer back into the separating funnel and repeat the process, using another portion of the aqueous solution — it is more effective to carry out solvent extraction twice with small portions of solvent rather than once with a large portion.

How do you decide which layer is the organic liquid?

One way to do this is to refer to the density — the more dense liquid will be the bottom layer. Alternatively, add some water and observe which layer increases in size — this will be the aqueous layer.

Knowledge check 15

Eucalyptus oil (density 0.922 g cm⁻³) can be produced by steam distilling eucalyptus leaves, washing the distillate with sodium chloride solution (density 1.22 g cm⁻³) and separating. Draw a labelled diagram of the apparatus used to separate the oil.

Tip

If a question asks about solvent extraction always check if the densities are given, and determine which is the organic layer.

Drying

The purified organic liquid may still contain water, and must be dried. The method for this is as follows:

1. Add a spatula of a drying agent, for example anhydrous calcium chloride or anhydrous magnesium sulfate, to the organic liquid in a beaker or conical flask.
2. Swirl.
3. Add more of the drying agent until the liquid changes from cloudy to clear.
4. Filter (Figure 16) or **decant** off the liquid into a clean, dry flask.

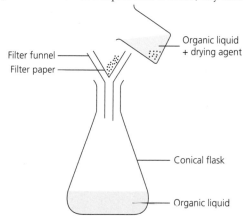

Figure 16 Gravity filtration diagram

Labels: Filter funnel, Filter paper, Organic liquid + drying agent, Conical flask, Organic liquid

Final distillation

The product can be redistilled. The boiling point measured during the distillation indicates the *identity* and the *purity* of the liquid product. If the boiling point is sharp then the product is pure, if it covers a range then impurities may be present. If the boiling point recorded during distillation is the same as that given in data books, then the correct product has been obtained.

If the boiling points of the reactants and the product are very similar, often fractional distillation (Figure 17) can be used instead of simple distillation.

In fractional distillation if the flask contains a mixture of liquids, the boiling liquid in the flask produces a vapour that is richer in the most volatile of the liquids present — the one with the lowest boiling point. Most of the vapour condenses in the column and runs back. As it does so it meets more of the rising vapour. Some of the vapour condenses and some of the liquid evaporates.

In this way, the mixture evaporates and condenses repeatedly as it rises up the column. But every time it does so, the vapour becomes richer in the most volatile liquid present. At the top of the column, the vapour can contain close to 100% of the most volatile liquid. So, during fractional distillation, the most volatile liquid with the lowest boiling point distils over first, then the liquid with the next lowest boiling point and so on. The boiling point can be measured as the liquid distils over during fractional distillation. Impurities increase the boiling point and the range over which the liquid boils.

Tip

A common error is to say 'decant off the anhydrous calcium chloride'. Be careful — at this stage the calcium chloride is no longer anhydrous, and it is the liquid that is decanted off.

Decant means to carefully pour a liquid from one container to another, in order to leave any solid in the bottom of the original container.

Knowledge check 16

Name a chemical used to dry eucalyptus oil.

Tip

If a liquid is pure it should distil over a narrow range, at the expected boiling point.

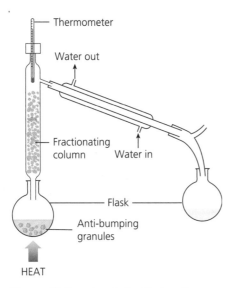

Figure 17 Fractional distillation diagram

Why is the percentage yield less than 100%?

There are theoretical and practical reasons for this.

Theoretical reasons

- Side reactions occur, so by-products may be produced instead of the expected products.
- The reagents used may be impure.
- The reaction is incomplete.

Practical reasons

- Some product is lost in the purification steps, for example in washing, separating in the separating funnel and in transfer between apparatus.
- Some product is lost in distillation.

Every organic liquid will be prepared using a slightly different method, so it is important that you understand the general preparation methods listed and apply them to different situations, as shown in the worked examples.

Worked example 1

The ester ethyl ethanoate was prepared from ethanol and ethanoic acid, using the method given below.

- Mix 12.0 cm^3 of ethanoic acid (density 1.05 g cm^{-3}) and 10.0 cm^3 of ethanol (density 0.789 g cm^{-3}) in a pear-shaped flask.
- Add 10 cm^3 of concentrated sulfuric acid slowly, with cooling and shaking.
- Add some anti-bumping granules and heat under reflux for 20 minutes.
- Distil off the ethyl ethanoate and collect the fraction between 74°C and 79°C.

→

Knowledge check 17

A common step in the purification of organic liquids is shaking with water in a separating funnel. Why is this not an appropriate method for purifying ethanoic acid?

- Place the crude ethyl ethanoate in a separating funnel and shake with sodium carbonate solution. Invert the funnel and open the tap occasionally.
- Allow the layers to separate and discard the lower aqueous layer.
- Add some calcium chloride solution to the ethyl ethanoate, to remove any ethanol impurities, shake, and run off the aqueous layer.
- Add a spatula of anhydrous calcium chloride and shake. Repeat until the ester is clear.
- Decant the liquid into a clean, dry, pear-shaped flask and redistill, collecting the ethyl ethanoate (density $0.92\,g\,cm^{-3}$) at the boiling point.

a Why must the concentrated sulfuric acid be added slowly and with cooling?

b Why is concentrated sulfuric acid used in this reaction?

c Write a balanced symbol equation for the reaction to prepare ethyl ethanoate.

d What is the nature and purpose of anti-bumping granules?

e How is the apparatus set up to heat under reflux?

f What is the function of the sodium carbonate solution?

g Why is the tap opened from time to time?

h Is the aqueous layer the top or bottom layer?

i Why is anhydrous calcium chloride added until the ester is clear?

j Why must the pear-shaped flask be clean and dry?

k Calculate the percentage yield if $8.2\,cm^3$ of ethyl ethanoate (density $0.92\,g\,cm^{-3}$) is collected.

Answer

a The concentrated sulfuric acid on dilution gives out a lot of heat. The slow addition with cooling dissipates the heat and avoids splashing, which would occur if the mixture became hot.

b It is a catalyst for the reaction. It is also a dehydrating agent and removes water, promoting the forward reaction.

c $CH_3CH_2OH + CH_3COOH \rightleftharpoons CH_3COOCH_2CH_3 + H_2O$

d Anti-bumping granules are small pieces of silica or broken, unglazed pottery. They promote smooth, even boiling.

e A condenser is fitted vertically to the pear-shaped flask.

f The distillate contains traces of unreacted ethanoic acid and concentrated sulfuric acid. The sodium carbonate solution removes these. (Sometimes the distillate is then washed with water, to remove traces of sodium carbonate solution.)

g The neutralisation of acid with sodium carbonate produces carbon dioxide gas; opening the tap releases this and avoids a build-up of pressure that might blow the stopper out of the funnel.

h The density of water and aqueous solutions is $1.0\,g\,cm^{-3}$. The density of the ester is $0.92\,g\,cm^{-3}$ so the aqueous layer is the bottom, denser layer.

i To remove water and dry the ester.

j The ester has just been purified and dried.

→

k density of ethanol $= \dfrac{\text{mass}}{\text{volume}}$ density of ethanoic acid $= \dfrac{\text{mass}}{\text{volume}}$

$$0.789 = \dfrac{\text{mass}}{10.0} \qquad\qquad 1.05 = \dfrac{\text{mass}}{12.0}$$

$$\text{mass} = 7.89\,\text{g} \qquad\qquad \text{mass} = 12.6\,\text{g}$$

$$\text{amount in moles} = \dfrac{\text{mass}}{M_r} \qquad \text{amount in moles} = \dfrac{\text{mass}}{M_r}$$

$$= \dfrac{7.89}{46.0} \qquad\qquad = \dfrac{12.6}{60.0}$$

$$= 0.17 \qquad\qquad = 0.21$$

The ethanoic acid is in excess.

moles of ethyl ethanoate $= 0.17 = \dfrac{\text{mass}}{M_r} = \dfrac{\text{mass}}{88.0}$

mass $= 0.17 \times 88.0 = 14.96\,\text{g}$

density of ethyl ethanoate $= 0.92 = \dfrac{14.96}{\text{volume}}$

volume $= \dfrac{14.96}{0.92} = 16.26\,\text{cm}^3$

% yield $= \dfrac{\text{actual yield}}{\text{theoretical yield}} \times 100$

$$= \dfrac{8.2}{16.26} \times 100 = 50.4\%$$

Knowledge check 18

Name a technique that could be used to boil an organic reaction mixture for 30 minutes.

Worked example 2

To prepare ethanal (boiling point 21°C) in the laboratory the following method was used:

- 7.5 g of sodium dichromate was placed into a distillation flask with 15 cm^3 of water.
- The apparatus was set up for distillation and a mixture of 3 cm^3 of concentrated sulfuric acid and 6 cm^3 of ethanol (density = 0.79 g cm^{-3}) placed into a tap funnel, and slowly added to the distillation flask.
- The flask was occasionally heated very gently until about 6 cm^3 of distillate was collected in a receiving flask surrounded by a mixture of ice and water.

a Identify two aspects of this preparation that might be hazardous. What steps would you take to minimise risks from these hazards?

b When the reaction is over, the remaining colour in the flask is orange. Explain which reactant is in excess.

c Why is distillation used instead of reflux in this experiment?

→

Tip

Remember that acidified potassium and sodium dichromate are oxidising agents and change colour from orange to green when warmed with a reducing agent such as ethanol.

d Suggest why the distillate is best collected in a flask surrounded by ice.

e If $2\,cm^3$ of the distillate containing ethanal, ethanol and water is added to a beaker of boiling water, explain what would happen.

f Describe how you would obtain a sample of ethanal from a mixture of ethanal, ethanol and water. Include in your answer a description of the apparatus you would use and how you would minimise the loss of ethanal. Your description of the apparatus can be either a description in words or a labelled sketch.

g What would be observed if the distillate was warmed with Tollens' reagent?

Answer

a The concentrated sulfuric acid is corrosive. It must be handled with care, and the handler should wear safety goggles and gloves.

Ethanol is flammable, so must be kept away from naked flames when measuring out. Consider using an electric heating mantle.

Use anti-bumping granules to prevent dangerous splashing and bumping.

b The orange colour indicates that the sodium dichromate is in excess. If it was all reduced then a green colour would be present.

c Distillation means that the ethanal is distilled off immediately and separated from the oxidising agent, hence further oxidation is prevented. Reflux would mean that the ethanal would still be in contact with the oxidising agent and fully oxidise to ethanoic acid.

d Ethanal has a low boiling point of 21°C, close to room temperature. The ice cools the flask and ensures that ethanal is collected as a liquid, and does not evaporate.

e Ethanal and ethanol can hydrogen bond with water — they are miscible with water. The liquids would boil, as both have boiling points less than 100°C.

f To separate these liquids the mixture could be distilled by placing the mixture in a flask, attaching a still head with thermometer and a water-cooled condenser from the side. An iced water-cooled receiving flask (to reduce evaporation of ethanal) should be used to collect the ethanal at 21°C (note that the temperature then rises and a different flask could be used to collect the ethanol at 78°C). An alternative answer would be a diagram similar to that in Figure 14, with the receiving flask cooled in iced water.

g A silver mirror, because the ethanal is oxidised.

Required practical 6

Tests for alcohol, aldehyde, alkene and carboxylic acid

Most organic functional groups have a characteristic qualitative test. For example, an alkene can be identified because it reacts with orange-brown bromine solution to produce a colourless solution. Functional group tests are usually carried out on a test-tube scale.

Table 9 summarises some of the qualitative tests that can be used to identify the presence of functional groups in organic compounds.

Tip

When drawing a distillation diagram make sure that the thermometer bulb is level with the outlet, the water direction is correct in the condenser, and it is an open system.

Tip

Organic compounds are often flammable so when carrying out analytical tests it is best to heat in a water bath, or use an electric heater.

Table 9

Test	Observation	Deduction
Shake with bromine water	Orange solution changes to a colourless solution	C=C present (alkene)
Warm with acidifed potassium dichromate(vi)	Orange solution changes to green solution	Primary/secondary alcohol or aldehyde
	Solution remains orange	Could be a tertiary alcohol (it is not oxidised)
Warm with Tollens' reagent (ammoniacal silver nitrate)	Silver mirror	Aldehyde
	Solution remains colourless	Ketone
Warm with Fehling's solution	Red ppt	Aldehyde
	Solution remains blue	Ketone
Warm with silver nitrate solution in ethanol	White ppt	Chloroalkane
	Cream ppt	Bromoalkane
	Yellow ppt	Iodoalkane
Add sodium carbonate or sodium hydrogen carbonate	Effervescence — the gas produced changes colourless limewater to cloudy	Carboxylic acid — the gas produced is carbon dioxide Note that phenol is weakly acidic and can react with bases and metals but it is not sufficiently acidic to react with carbonates
Add magnesium	Effervescence — a pop is heard when a lighted splint is applied to the gas produced	Carboxylic acid — the gas produced is hydrogen
Warm with ethanol and a few drops of concentrated sulfuric acid; cautiously smell	Sweet smell	Ester produced — the organic substance could be a carboxylic acid

Tip

The lack of a hydrogen atom bonded to the carbon with −OH bonded to it makes tertiary alcohols resistant to oxidation by acidified dichromate(vi).

Tip

Primary alcohols and aldehydes are oxidised to carboxylic acids by acidified potassium dichromate(vi). Secondary alcohols are oxidised to ketones by acidified potassium dichromate(vi).

Knowledge check 19

Why does bromine water decolourise when added to an alkene?

Knowledge check 20

A silver mirror is formed when Tollens' reagent is warmed with ethanol. Which metal ion is reduced in this reaction? Write an ionic equation for the reduction.

Knowledge check 21

Write balanced symbol equations for the reaction of propanoic acid with magnesium and with sodium carbonate, and give observations.

Worked example 1

The following tests were carried out on a compound A and the observations recorded.

Test	Observation
1 Add a spatula of sodium carbonate to A	No bubbles produced
2 Warm with a few drops of acidified potassium dichromate solution	Orange solution changes to green
3 Warm a few drops of A with Tollens' reagent	Silver mirror

Make deductions based on these observations. What type of organic chemical is A?

Answer

In test 1 A does not react to produce bubbles of carbon dioxide so A is not a carboxylic acid. Test 2 shows that A has been oxidised so it could be a primary alcohol, secondary alcohol or an aldehyde. Test 3 shows that A is an aldehyde.

Worked example 2

A student warms some propan-1-ol with acidified potassium dichromate(VI) using the apparatus shown in Figure 18.

Figure 18 Heating propan-1-ol with acidified potassium dichromate(VI)

a State one reason why this apparatus is not suitable to use in this experiment.

b Describe a more suitable way of carrying out this test.

c Name the organic compound produced.

d What is observed in the test tube?

e If this reaction was carried out by heating for an extended time to produce a final product, suggest a suitable way of carrying out the reaction.

Answer

a Propan-1-ol is flammable.

b Warm the boiling tube gently in a water bath.

c Propanoic acid — if the alcohol is not fully oxidised then there may also be some propanal.

d Orange solution changes to green solution.

e Reflux with some anti-bumping granules added to the mixture.

Required practical 7a

Measuring the rate of a reaction by a continuous monitoring method

To investigate the rate of a reaction continuous monitoring can be used. In this method, the reaction is monitored throughout its course and the amount of reactant or product is established at different time intervals throughout the reaction. A variety of methods can be used to follow the progress of the reaction against time: most measure a change in the amount or concentration of a reactant or product during the reaction.

To decide on a monitoring method you need to examine the equation of the reaction and identify a reactant or a product that can be measured. Examples include:

- a gaseous product, which can be monitored by measuring *volume of gas produced* or by *loss in mass* of the reaction system
- a coloured reactant or product, which can be monitored using *colorimetry*
- a titratable reactant or product, which can be monitored by *sampling, quenching* and *titrating*
- a directly measurable reactant or product, i.e. H^+ ions or OH^- ions, by measuring pH using a *pH meter*

Measuring gas volume

For gases that are not very soluble in water, such as oxygen or hydrogen, the gas may be collected under water in an inverted measuring cylinder (Figure 19) or burette.

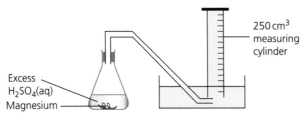

Figure 19 Collection and measurement of a gas under water

If a gas is soluble in water (e.g. carbon dioxide) then a gas syringe may be attached to a sealed reaction vessel to measure the volume of gas produced (Figure 20).

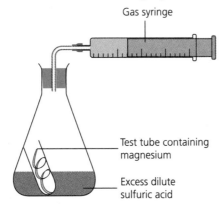

Figure 20 Using a gas syringe to collect gas

To start the reaction magnesium is dropped into the flask and the stopper quickly replaced. There are some errors in this method:

- Some gas may escape due to the time lag between adding the magnesium and replacing the bung, resulting in a decrease in the measured volume of gas.

  To reduce this error, the magnesium could be suspended by a string above the acid and the stopper loosened just enough to release the thread, dropping the magnesium into the acid. Alternatively, the magnesium could be placed in a small tube in the conical flask, acid added, bung replaced and then the flask swirled to mix the reactants.

- When the stopper is replaced, the volume of the bung displaces the same volume of air into the measuring cylinder, increasing the volume. Again the alternative methods above reduce this error.

Tip

Note that if this experiment was carried out using magnesium carbonate and acid, carbon dioxide gas would be produced. A different error may occur in this case — some of the gas produced may dissolve in water. To reduce this error, use a gas syringe (Figure 20).

Tip

Note that most gas syringes hold $100\,cm^3$ of gas, so before carrying out an experiment you need to check by moles calculation that the mass of solid used produces less than $100\,cm^3$ of gas. The example above, with $0.12\,g$ of magnesium, produces $120\,cm^3$ of gas, so the mass of magnesium must be halved before using a gas syringe, and a more accurate balance used.

In this experiment 0.12 g of magnesium was reacted with excess acid and 120 cm³ of hydrogen gas was produced. These values can be used to calculate the relative atomic mass of magnesium (A_r). (All volumes were recorded at room temperature and pressure.)

$$\text{amount in moles of } H_2 = \frac{\text{volume } H_2}{24\,000} = \frac{120}{24\,000} = 0.005\,\text{mol}$$

$$Mg + H_2SO_4 \rightarrow MgSO_4 + H_2$$

The ratio is $1:1$, so moles Mg = 0.005 mol.

$$A_r\,Mg = \frac{\text{mass}}{\text{moles}} = \frac{0.12}{0.005} = 24.0$$

Therefore:

$$A_r(Mg) = 24.0$$

Many experiments involving the production of a gas involve heating a solid (e.g. decomposition of a carbonate). If a solid is heated to generate a gas, a test tube or boiling tube should be used as shown in Figure 21, rather than a conical flask. Remember that the volume of a gas depends on the temperature and pressure. An increase in temperature causes a gas to expand significantly and this could cause an error in this type of experiment.

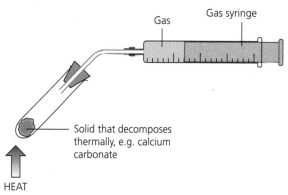

Figure 21 Heating a sample and collecting the gas; note that the reaction is over when the gas syringe stops moving

To investigate the rate the volume of gas is measured against time and a graph of gas volume against time is plotted.

The gas in a syringe will always contain some air pushed into the syringe by the gas that is generated in the reaction. This will not affect the result because the total volume of gas and air collected in the syringe will be identical to the volume of gas produced.

Measuring change in mass

A reaction in which a gas is produced can also be monitored by measuring the mass over a period of time. This method is not suitable for hydrogen, which has a very low formula mass, so the mass lost would be small and very difficult to measure. The apparatus is shown in Figure 22. A graph of mass against time is often plotted.

If a reaction is very exothermic it is difficult to accurately determine the rate because the higher temperature causes an increase in reaction rate, which causes the reaction to proceed faster.

Knowledge check 22

In this experiment two measurements were taken: mass of magnesium = 0.12 g and volume of hydrogen = 120 cm³. Calculate the percentage error in both readings if the balance uncertainty is ±0.01 g and the measuring cylinder uncertainty is ±1 cm³.

Knowledge check 23

In an experiment a gas could be collected in a 100 cm³ measuring cylinder, a 100 cm³ gas syringe or a 50.0 cm³ burette. Which measuring instrument has higher resolution?

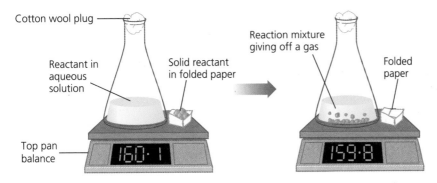

Figure 22 Following the course of a reaction by measuring the change of mass

Measuring change in a reactant or product by titration

Sometimes samples of the reacting mixture are taken at various times during the course of a reaction and the reaction is **quenched** — this means that the reaction is stopped (Figure 23). Methods of quenching include rapid cooling, adding a chemical to remove a reactant that is not being monitored, or adding a large known volume of water to the sample. Each sample may then be titrated to find the concentration of the reactant or product and a graph of concentration against time drawn.

Tip

The shape of the graph and the calculation of half-life can be used to determine the order. The gradient of tangents at different concentrations can give the rate and a rate against concentration graph also be used to find the order.

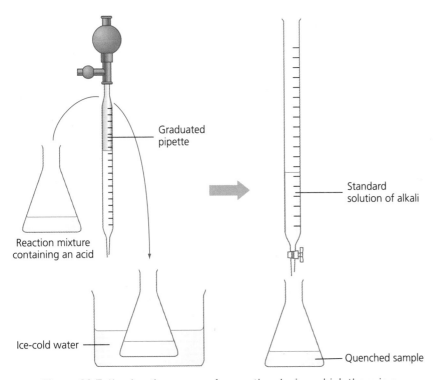

Figure 23 Following the course of a reaction during which there is a change in concentration of acids present by removing measured samples of the mixture at intervals, quenching and then determining the concentration of one reactant or product by titration

Measuring a coloured reactant or product

Measurement of a coloured reactant or product can be used for continuous rate monitoring. A **colorimeter** measures the colour intensity of a solution. A calibration curve should be set up first with known concentrations of the reactant or product so that colorimeter readings are related to concentration.

In the continuous rate method one experiment is carried out and the colorimeter readings are converted to concentration using the calibration curve. A graph of concentration against time is drawn and the shape of this curve can give the order with respect to the coloured reactant.

Worked example 1

Suggest a suitable method for measuring the rate of each of these reactions:

a $CH_3COOCH_3(l) + H_2O(l) \rightarrow CH_3COOH(aq) + CH_3OH(aq)$
b $C_4H_9Br(l) + H_2O(l) \rightarrow C_4H_9OH(l) + H^+(aq) + Br^-(aq)$
c $MgCO_3(s) + 2HCl(aq) \rightarrow MgCl_2(aq) + CO_2(g) + H_2O(l)$

Answer

a This is hydrolysis of an ester and there will be a change in concentration of ethanoic acid as it is produced in the reaction. Remove samples at intervals, quench the reaction by cooling and then titrate against alkali to determine the concentration of acid at different times.

b Use a pH meter to measure the changes in conductivity as the number of hydrogen ions increases.

c A gas — carbon dioxide — is produced. Carry out the reaction in a flask, with a loose plug of cotton wool in its neck, on a balance and record the loss in mass at regular intervals. Alternatively, measure the volume of carbon dioxide in a gas syringe at regular intervals against time.

Knowledge check 24

Magnesium reacts with copper(II) sulfate. What method could be used to determine the effect of the concentration of copper(II) sulfate on the rate of reaction?

Finding the order of reaction

It is possible to find the order for a particular reagent by the continuous monitoring method. This method works if the reaction involves only one reagent. Alternatively, if more reagents are present the reaction can be set up so that the reaction rate depends solely on one reagent, and all the other reagents are present in such a *large excess* that their concentration is *nearly constant* and so they have little effect on the reaction rate. The order can be determined from the shape of the concentration–time graph (Figure 24).

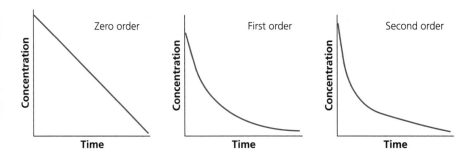

Figure 24 Concentration–time graphs

A zero-order reaction has a straight-line graph. For a first-order graph the graph of concentration against time is a curve. From this graph the **half-life** of the reaction can be worked out. If the *half-life is constant* then the reaction is *first order*.

Figure 25 shows a concentration–time graph. From the graph you can see that:
- time taken for concentration to fall by half from 1.0 to 0.5 is 100 s
- time taken for concentration to fall by half from 0.5 to 0.25 is 100 s
- time taken for concentration to fall by half from 0.25 is 0.125 is 100 s

Hence the half-life is constant, at 100 s, proving that the graph is for a first-order reaction.

The **half-life** of a reaction is the time taken for the concentration of one of the reactants to fall by half.

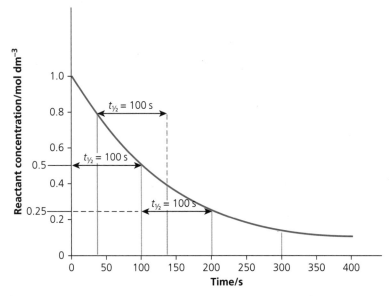

Figure 25 A constant half-life shows that the reaction is first order

The graph of concentration against time for a second-order reaction is shown in Figure 26. The half-life for a second-order reaction is not constant. Take the gradients of tangents at points along the curve to find the rate.

Tip

If the half-lives show that a reaction is not first order, it may be second order but this needs to be checked by finding the rate at different points. Some reactions may have a fractional order.

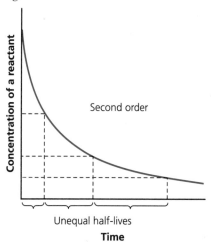

Figure 26 A graph of concentration against time for a second-order reaction

Another way of determining order is to find the rate at different concentrations by finding the gradients of tangents at these points and plotting a rate against concentration graph. The shape gives the order (Figure 27).

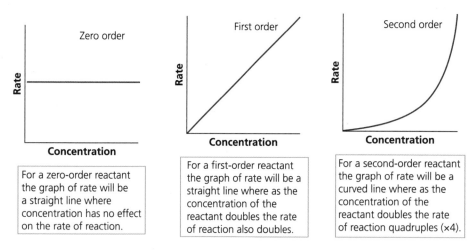

For a zero-order reactant the graph of rate will be a straight line where concentration has no effect on the rate of reaction.

For a first-order reactant the graph of rate will be a straight line where as the concentration of the reactant doubles the rate of reaction also doubles.

For a second-order reactant the graph of rate will be a curved line where as the concentration of the reactant doubles the rate of reaction quadruples (×4).

Figure 27 Rate against concentration graphs

Worked example 2

Calcium carbonate reacts with hydrochloric acid. In this reaction carbon dioxide is produced. If the calcium carbonate is present in excess it is possible to investigate the rate of this reaction and determine the order with respect to the acid. The rate of reaction can be assessed by monitoring the loss in mass. A conical flask containing reactants, with a cotton wool plug in the neck, is placed on a balance and the mass is recorded every 10 seconds initially and then every 20 seconds.

a What is the purpose of the cotton wool plug?

b What is observed in the flask during the reaction?

c Why is the time interval of recording changed?

d When do you stop taking readings in this experiment?

e Suggest why this reaction does not go to completion in the time available for the experiment.

f Write a balanced symbol equation, with state symbols, for the reaction that occurs in this experiment.

g Another method of monitoring the rate of this reaction is by gas collection. The gas can be collected under water in a inverted measuring cylinder. State a source of error in this experiment.

h Collecting the gas in a $100\,cm^3$ gas syringe is a more effective method of gas collection. Show by calculation that a $100\,cm^3$ gas syringe is suitable to collect the gas produced by the reaction of $10\,g$ (excess) of calcium carbonate with $15.0\,cm^3$ of $0.5\,mol\,dm^{-3}$ HCl(aq) at room temperature and pressure.

Answer

a To prevent loss of acid spray.

b Bubbles; the calcium carbonate disappears and a colourless solution is formed.

c The reaction rate is fastest at the beginning and so it is best to record readings more often.

d When two readings are constant, showing that no more carbon dioxide is being lost and the reaction is over.

e The HCl(aq) is used up in the reaction and so the concentration of HCl(aq) decreases so much that the reaction becomes so slow that it seems to stop, even though it may not have gone to completion.

f $CaCO_3(s) + 2HCl(aq) \rightarrow CaCl_2(aq) + H_2O(l) + CO_2(g)$

g Carbon dioxide may dissolve in the water.

h moles of HCl $= \dfrac{\text{volume (cm}^3) \times \text{concentration}}{1000} = \dfrac{15.0 \times 0.5}{1000} = 0.0075$

ratio $= 2\,mol\,HCl : 1\,mol\,CO_2$

moles $CO_2 = 0.00375 = \dfrac{\text{vol (cm}^3)}{24\,000}$

volume $= 90\,cm^3$

Knowledge check 25

What is meant by the term half-life?

Required practical 7b

Measuring the rate of a reaction by an initial rate method

In this method the rate is determined immediately after the start of the reaction. At this point all of the concentrations are known. A series of experiments is carried out in which the initial concentration of one of the reactants is varied systematically while the concentration of the other reactants is kept constant. The time, t, is measured to a fixed point in the reaction. In the method of initial rates it is necessary to determine the initial rate for each experiment. Often results are plotted on a concentration–time graph. Initial rates can be determined from the concentration–time graph. A tangent is drawn at $t = 0$ and the rate is the gradient of this tangent.

Tangents to a curve

The word tangent means 'touching' in Latin. The tangent is a straight line that just touches the curve at a given point and does not cross the curve. Figure 28 shows how to draw a tangent at a point (x, y).

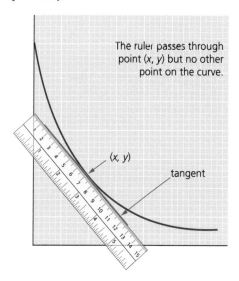

The ruler passes through point (x, y) but no other point on the curve.

(x, y)

tangent

Figure 28 Drawing a tangent to a curve

1 Place your ruler through the point (x, y).

2 Make sure your ruler goes through the point and does not touch the curve at any other point.

3 Draw a ruled pencil line passing through the point (x, y).

To calculate the *gradient* of a curve at a particular point it is necessary to draw a tangent to the curve at the point and calculate the gradient of the tangent.

Knowledge check 26

For the rate equation:

$$\text{rate} = k[C][D]^2$$

state the order of reaction with respect to C and D and state the units of the rate constant.

Knowledge check 27

What are the units of the gradient for a graph of rate against concentration?

Worked example 1

A student recorded the total volume of gas collected in a reaction at 20-second intervals and plotted the graph in Figure 29. Use the graph to calculate the rate of reaction at 60 s. State the units.

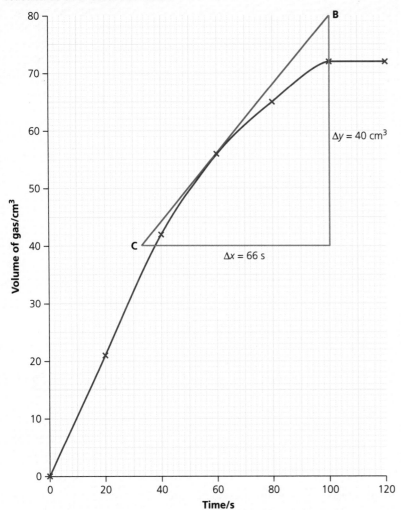

Figure 29 Volume of gas against time graph

Answer

First draw the tangent at time 60 s — this is the red line on the graph. The rate at this point is given by the gradient of this tangent. To find the gradient choose two points, B and C, far apart on the line and form a triangle, as shown in green.

$$\text{gradient } (m) = \frac{\text{change in } y\text{-axis}}{\text{change in } x\text{-axis}} = \frac{\Delta y}{\Delta x} = \frac{40}{66} = 0.61 \text{ (2 s.f.)}$$

$$\text{units} = \frac{\text{cm}^3}{\text{s}} = \text{cm}^3 \text{s}^{-1}$$

To find the slope of the curve at any other point, draw a tangent line at that point and then determine the slope of that tangent line.

If the concentrations of the reagents are known at the start of the reaction, a measurement of the initial gradient ($t = 0$) gives a numerical value of the rate at these concentrations.

If the experiment is repeated with the concentration of one of the reactants doubled, while the other concentrations are kept the same, a new initial rate can be established. Therefore, it is possible to see the effect of this reactant on the overall rate. If it is found that the gradient has doubled, then the reaction is first order with respect to that reactant. If, however, the rate increases four fold, it would be second order with respect to the reagent. Changing the concentrations of each reactant in turn allows the order of reaction with respect to each reagent to be determined.

Clock reactions

To carry out several full experiments using the methods described on page 37–40 to provide data for the initial rate of a reaction is time consuming. Instead an approximation to obtain an initial rate is used.

The time taken to reach a specific point in the reaction soon after the reaction has started can be recorded and the reaction repeated with different concentrations to determine how the time taken to reach this point changes. The rate to this point is taken to be proportional to 1/time ($1/t$) for each reaction.

Clock reactions are ideal for measuring the time for a certain amount of product to be formed — a clock reaction measures the time from the start of the reaction until there is a visual change (preferably sudden), such as:

- appearance of a precipitate
- disappearance of a solid
- change in colour

The reaction of hydrogen peroxide with iodide ions in acid solution can be set up as a clock reaction:

$$H_2O_2(aq) + 2H^+(aq) + 2I^-(aq) \rightarrow I_2(aq) + 2H_2O(l)$$

A small, known amount of sodium thiosulfate ions is added to the reaction mixture, which also contains starch indicator. At first the thiosulfate ions react with any iodine, I_2, as soon as it is formed, turning it back to iodide ions, so there is no colour change. At the instant when all the thiosulfate ions have been used up, free iodine is produced and this immediately gives a deep blue-black colour with the starch. The reaction time t for the initial part of the reaction to take place is measured and $1/t$ calculated as a measure of the initial rate of the reaction. The experiment can be repeated with different initial concentrations of the reactants and the time taken to reach the blue-black colour recorded.

Tip

It is assumed that the reaction proceeds at a constant rate to the point that it is measured. This is not strictly true because, as soon as the reaction starts, the rate begins to slow. However, the approximation is good enough to determine an integer order.

Knowledge check 28

In a reaction, doubling the concentration of hydrogen doubles the rate of reaction when the concentration of other reactants is constant. Tripling the concentration of NO gas increases the rate by a factor of nine, when other reactants have constant concentration. Write the rate equation for the reaction.

Worked example 2

The graph shown in Figure 30 is for an iodine clock reaction. The blue-black colour appears at 60 seconds. Explain, using the graph, why the estimate of the initial rate of a reaction determined by this clock reaction is close to, but not equal to, the true initial rate.

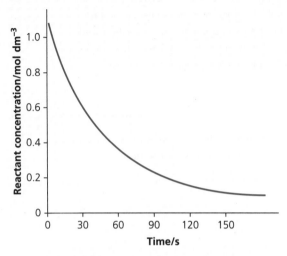

Figure 30 A concentration–time graph

Answer

The graph gradient decreases, showing that the reaction proceeds quickly at the start and slows down towards the finish, so the rate when the clock reaction ends is slower than the initial rate. A tangent to the curve is drawn at $t = 0$ and at $t = 60$ seconds (blue lines in Figure 31), and the gradient calculated.

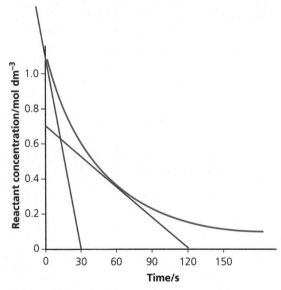

Figure 31 Drawing tangents at different times to find the rate

From the graph:

- the initial rate, at time $t = 0$ seconds, is $\dfrac{1.07}{30} = 0.036 \, mol \, dm^{-3} \, s^{-1}$
- the rate after 60 seconds is $\dfrac{0.7}{120} = 0.006 \, mol \, dm^{-3} \, s^{-1}$

These values show that the rate slows considerably as the reaction progresses and at time 60 seconds, when the colour change of the clock reaction occurs, the rate is much slower as the reactants have started to be used up. The rate is at its initial maximum before the reactants have started to be used up. Hence the clock reaction is an approximation and works best if the reaction has not progressed very far.

Worked example 3

The iodate(v) ion is an oxidising agent and reacts with sulfate(IV) ions in acidic solution to produce iodine in solution, according to the following equations:

Stage 1: $IO_3^- + 3SO_3^{2-} \rightarrow I^- + 3SO_4^{2-}$

Stage 2: $IO_3^- + 5I^- + 6H^+ \rightarrow 3H_2O + 3I_2$

Overall: $2IO_3^- + 5SO_3^{2-} + 2H^+ \rightarrow I_2 + 5SO_4^{2-} + H_2O$

Iodine is only liberated when acid is added. It is possible to determine the effect of changing the concentration of acid on the initial rate of this reaction by timing how long it takes for iodine to be produced.

a What should be added to the reaction mixture to indicate when iodine has been produced?

b The reactants were placed in a beaker and the timer started, and the contents stirred with a glass rod. Why were the contents stirred?

c When is the timer stopped?

d The concentration of acid was changed in this experiment. Name two controlled variables in the experiment.

e The experiment was designed to determine the order of reaction with respect to hydrogen ions in the reaction. What changes would you make to the experiment so that the order of reaction with respect to iodate(v) ions could be determined?

f The reaction is first order with respect to hydrogen ions. In an experiment to determine the order of this reaction a value of 0.963 was obtained. Calculate the percentage error in this result.

g The experimental error resulting from the use of the apparatus was determined to be 2.1%. Explain what this means in relation to the practical technique used.

Answer

a Starch

b To ensure that the reactants all mixed and reacted.

→

c When the blue-black colour indicating starch appears.

d Volume of all solutions/concentrations of the other solutions/temperature

e Keep the concentration of acid constant and vary the concentration of iodate(v).

f error = 1 − 0.963 = 0.037

$$\% \text{ error} = \frac{0.037}{1} \times 100 = 3.7\%$$

g The result error is greater than the apparatus error so the fault is due to the person carrying out the experiment.

In a clock experiment reacting sulfuric acid, potassium iodate(v) and sodium sulfate(IV) solutions, in the presence of starch, all of the concentrations were kept constant except the concentration of acid, hence the rate equation for the reaction is:

rate = $k[IO_3^-]^x[SO_4^{2-}]^y[H^+]^z$

This simplifies to:

rate = $k[H^+]^z$

where z is the order of reaction with respect to hydrogen ion concentration. This expression can be written:

log(rate) = $z \log[H^+]$ + constant

The event being measured is fixed, i.e. the first appearance of the blue–black colour, so it is possible for the rate to be expressed as:

rate = $\frac{1}{t}$

The total volume of solution in each experiment is constant and so the [H⁺] can be represented by the volume of sulfuric acid. Therefore, the rate expression becomes:

$\log\left(\frac{1}{t}\right)$ = $z \log$ (volume of sulfuric acid) + constant

The results are given in Table 10. 1/t has been calculated as a measure of the rate and log 1/t and log(volume of sulfuric acid) are also calculated.

Tip

You do not need to be able to deduce this expression, but you should be able to calculate logs, as shown in Table 10.

Table 10

Volume of acid/cm³	25	35	45	55	70	85
t/s	42	28	10	15	11	9
$\frac{1}{t}/s^{-1}$	0.0238	0.0357	0.500	0.677	0.0909	0.111
$\text{Log}\frac{1}{t}$	−1.62	−1.40	−1.30	−1.18	−1.04	−0.96
Log (volume)	1.40	1.54	1.65	1.74	1.85	1.93

A graph of of $\log(1/t)$ (y-axis) against $\log V$ (volume) (x-axis) is shown in Figure 32. It is a straight line of gradient z, and from this the order with respect to hydrogen ion concentration can be determined. The graph is of the type $y = mx + c$ and therefore the gradient m will give a value for the order, z.

The gradient is $0.60/0.50 = 1.2$, so the reaction is first order with respect to hydrogen ion concentration.

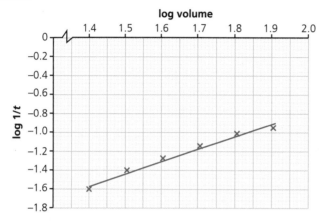

Figure 32 Graph of $\log 1/t$ against $\log$(volume)

Required practical 8

Measuring the EMF of an electrochemical cell

Electrochemical cells produce an electric potential difference (voltage) from a redox reaction. In an electrochemical cell the two half-reactions occur in separate half-cells. The electrons flow from one cell to the other through a wire connecting the electrodes.

The potential difference between the two half-cells is *maximum* when no current is flowing and is called the cell potential (or electromotive force). The cell potential can be measured using a high-resistance voltmeter. As very little current is drawn by the voltmeter, each electrode is effectively in equilibrium, and the measured cell potential will be close to the standard cell potential (assuming standard conditions are used).

In the laboratory many different cells can be set up using strips of metals dipping into solutions of their own ions connected by a high-resistance voltmeter and a salt bridge. If the half-cell is for a system that contains two ions (e.g. Fe^{2+} and Fe^{3+}) a carbon or platinum electrode is used.

Tip

Standard conditions for electrochemical measurements are: temperature of 298 K, gases at a pressure of 100 kPa, solutions at a concentration of 1.0 mol dm^{-3}.

What is a salt bridge?

The electric circuit is completed by a salt bridge connecting the two solutions. It allows ions to flow while preventing the solutions from mixing. The charge of all ions (cations and anions) in both half cells must remain zero to maintain the electron flow, and the salt bridge allows ions to move between half-cells and keep the charge in each container zero.

Often a salt bridge is a piece of filter paper soaked in potassium nitrate solution. Potassium nitrate is used because all potassium salts and nitrate salts are soluble, so the potassium nitrate does not react to produce precipitates with any of the ions in the half-cells

Tip

Cell potentials (EMF) are at a maximum when no current flows because under these conditions no energy is lost due to the internal resistance of the cell as the current flows.

Knowledge check 29

State the conditions required to measure standard electrode potentials.

Worked example

In an experiment the cell in Figure 33 was set up with copper as the negative electrode. The cell potential was altered by changing the concentration of the iron(II) ions in the solution to change the position of equilibrium in the half-cell, and the results recorded.

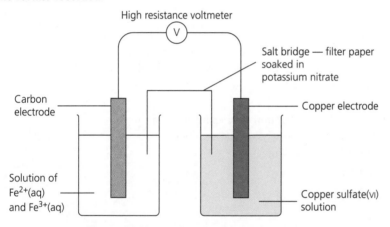

Figure 33 Measuring electrode potentials

The results for five different solutions, labelled 1–5, each containing a different ratio of iron(II) ions to iron(III) ions, are shown below.

Solution	Ratio $Fe^{2+}:Fe^{3+}$	Voltage/V
1	1:5	0.574
2	1:3	0.562
3	1:1	0.534
4	3:1	0.506
5	5:1	0.495

Knowledge check 30

Write two ionic half-equations and the overall balanced equation for each of the following redox reactions. In each example state which atom, ion or molecule is oxidised and which is reduced:

a magnesium metal with copper(II) sulfate solution

b aqueous chlorine with a solution of potassium bromide

The electrode potentials of the two half-cells were:

$$Cu^{2+} + 2e^- \rightleftharpoons Cu(s) \qquad E^\ominus = +0.34\,V$$

$$Fe^{3+} + e^- \rightleftharpoons Fe^{2+}(aq) \qquad E^\ominus = +0.77\,V$$

a Outline a method that could be used to identify the positive electrode in an electrochemical cell.

b Why does the iron half-cell use a carbon electrode?

c Write the overall equation for the redox reaction that occurs.

d State the direction of electron flow in the external circuit.

e Calculate the electrode potential of the cell.

f Describe the function of the salt bridge.

g Explain why a piece of wire is not used instead of a salt bridge.

h State the trend in concentration of iron(III) ions in the mixture from solution 1 to solution 5.

i Using Le Chatelier's principle, explain the effect of the trend in question (h) on the position of equilibrium for the iron half-cell reaction and on the electrode potential.

Answer

a When the cell is set up, if the reading on the voltmeter is positive then the metal connected to the positive terminal on the voltmeter is the positive electrode; if the reading is negative then the metal connected to the negative terminal is the positive electrode.

b The iron half-cell uses a graphite electrode, as both iron species involved in the redox reaction are ionic and in solution. The graphite is a good conductor, has low reactivity and so does not react with the solutions.

c In the copper half-cell the reaction that occurs is:

$$Cu(s) \rightarrow Cu^{2+} + 2e^- \qquad \text{(oxidation)}$$

In the iron half-cell, the reaction that occurs is:

$$Fe^{3+} + e^- \rightarrow Fe^{2+}(aq) \qquad \text{(reduction)}$$

Hence, the overall equation for the reaction occurring is:

$$Cu(s) + 2Fe^{3+}(aq) \rightarrow Cu^{2+}(aq) + 2Fe^{2+}(aq)$$

d The copper atoms form ions and give up electrons, which flow from the copper half-cell through the external circuit to the iron half-cell. This is from right to left in the diagram.

→

Tip

In a cell oxidation occurs in one half-cell and electrons are given up, which build up on the metal electrode. It is the negative electrode. The electrons flow out of this electrode through the circuit towards the positive electrode.

Tip

At the positive electrode, ions take electrons from the metal electrode and form atoms. The electrode is positive because it loses electrons.

e $Cu^{2+} + 2e^- \rightleftharpoons Cu(s)$ $E^\ominus = +0.34\,V$

$Fe^{3+} + e^- \rightleftharpoons Fe^{2+}(aq)$ $E^\ominus = +0.77\,V$

$E^\ominus_{cell} = +0.77 - 0.34 = +0.43\,V$

f   The salt bridge is needed to complete the electrical circuit of the electrochemical cell, while keeping the solutions separated. As the concentration of copper ions increases, nitrate ions enter the solution from the salt bridge to balance the charge. As the concentration of iron(III) ions decreases, potassium ions enter the solution from the salt bridge to balance the charge.

g   In a wire the conducting species are electrons, but in the half-cell the conducting species are ions. As a result no current could flow between the half-cells using a wire, hence the circuit would not be complete and the cell potential could not be measured.

h   Moving from solution 1 to solution 5 the concentration of iron(III) ions decreases.

i   To oppose the decrease in iron(III) concentration the position of the equilibrium shifts to the left-hand side of the equation.

$Fe^{3+} + e^- \rightleftharpoons Fe^{2+}$

This decrease in iron(III) ion solution concentration results in less positive voltages.

The cell potential is calculated as:

$E_{cell} = E_{\text{reduction reaction}} - E_{\text{oxidation reaction}}$

Because the electrode potential of the iron reaction becomes more negative with decreasing concentration, the difference between the electrode potential of the iron half-cell and the copper half-cell gets smaller and so the voltage reading gets smaller.

> **Tip**
>
> The further to the left the position of equilibrium is, the less positive (more negative) the value of the electrode potential.

Required practical 9

Investigate how pH changes when a weak acid reacts with a strong base and when a strong acid reacts with a weak base

The pH of solutions can be measured with a pH meter, pH probe, data logger or narrow-range pH paper. The pH change in a solution during an acid–base titration can be measured using a pH probe. When using a pH meter or probe, it is necessary to calibrate it before use, so that accurate pH values can be obtained.

Periodic measurement of the pH of the reaction solution allows plotting of a graph of pH against volume of solution added. Such a graph is called a pH titration curve and each has a characteristic shape for the various combinations of strong and weak acids and bases (Figure 34).

Skills Guidance

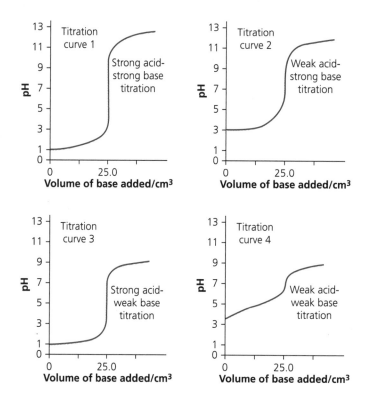

Figure 34 Titration curves

Worked example

a Methanoic acid is a weak acid. In an experiment, a calibrated pH meter was used to measure the pH of methanoic acid solution. At 20°C the pH of a $0.100 \, \text{mol dm}^{-3}$ solution was 2.37.

  i Explain why a pH meter should be calibrated before use.

  ii Explain how a pH meter is calibrated.

  iii Write an equation for the dissociation of methanoic acid and explain what is meant by *weak acid*.

  iv Write an expression for the equilibrium constant, K_a, for the dissociation of methanoic acid in aqueous solution and calculate the value of K_a for this dissociation at 20°C. Give your answer to the appropriate precision.

b A student used aqueous sodium hydroxide to determine the titration curve for the reaction of methanoic acid and sodium hydroxide solution. $25.0 \, \text{cm}^3$ of $1.50 \times 10^{-2} \, \text{mol dm}^{-3}$ methanoic acid was placed in a beaker at 25°C. The sodium hydroxide was added in $2.0 \, \text{cm}^3$ portions from a burette and the pH of the reaction mixture was measured using a pH meter.

  i Describe how $25.0 \, \text{cm}^3$ of methanoic acid was accurately measured and placed in the conical flask.

  ii Describe how the burette was prepared.

  iii Write a balanced symbol equation for the reaction between HCOOH and NaOH.

iv Why was the reaction mixture swirled after the addition of each $2.0\,cm^3$ portion of sodium hydroxide?

v The pH curve for this titration is shown in Figure 35. Calculate the value of the concentration in $mol\,dm^{-3}$ of the aqueous sodium hydroxide.

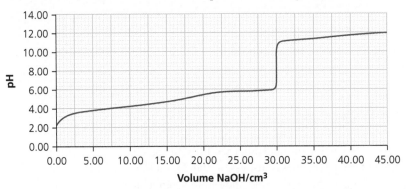

Figure 35 A titration curve

vi The pH ranges in which the colour changes for three acid–base indicators are shown below. Explain which one of the three indicators is suitable for this titration.

Indicator	pH range
Metacresol purple	7.40–9.00
2,4,6-trinitrotoluene	11.50–13.00
Ethyl orange	2.4–4.8

Answer

a i After storage, a pH meter does not give accurate readings because the glass electrode in the pH meter does not give a reproducible EMF over longer periods of time.

ii Calibration should be performed with at least two standard buffer solutions that span the range of pH values to be measured. The pH probe is rinsed thoroughly with deionised water, and shaken to remove excess water. It is then placed in a standard buffer solution of one pH, ensuring the bulb is fully immersed and allowed to sit until the pH stabilises. The reading is adjusted to the pH of the buffer. The probe is then rinsed with deionised water and the process repeated with a different pH buffer.

iii $HCOOH \rightleftharpoons HCOO^- + H^+$

The reversible arrow shows that the dissociation is incomplete because HCOOH is a weak acid.

Tip

An alternative technique for calibrating is to record the pH reading for each buffer and plot a graph of the pH of the recorded pH (x-axis) against the pH of the buffer solution. This calibration curve can be used to convert pH readings in an experiment into more accurate values.

iv $K_a = \dfrac{[HCOO^-][H^+]}{[HCOOH]}$

$pH = -\log[H^+]$

$2.37 = -\log[H^+]$

$[H^+] = 4.27 \times 10^{-3}$

$K_a = \dfrac{(4.27 \times 10^{-3})^2}{0.100}$

$K_a = 1.82 \times 10^{-4}$

The answer is given to three significant figures, as that is the accuracy of the data given.

b i A 25.0 cm³ pipette was rinsed with methanoic acid and then filled using a safety pipette filler, until the bottom of the meniscus was on the 25.0 cm³ mark at eye level. It was transferred to the beaker and allowed to run out. The pipette was touched on the surface of the liquid to expel the last drops.

ii The burette was rinsed with sodium hydroxide solution, and then filled to above the zero mark. The solution was run down until the bottom of the meniscus was on the zero mark at eye level, to ensure that the tap was filled. The burette was checked to ensure that there were no bubbles.

iii $NaOH + HCOOH \rightarrow HCOONa + H_2O$

iv To ensure that the solution is homogeneous and the pH is uniform throughout.

v moles of methanoic acid $= \dfrac{25.0 \times 1.5 \times 10^{-2}}{1000} = 3.75 \times 10^{-4}$

ratio $= 1:1$

moles NaOH $= 3.75 \times 10^{-4} = \dfrac{30.0 \times \text{concentration}}{1000}$

conc. $= 0.0125 \,\text{mol}\,\text{dm}^{-3}$

vi The pH change at the end point (vertical portion) is between 6 and 11. Metacresol purple changes colour within this range and is suitable to use.

Required practical 10

Preparation of a pure organic solid (and test of its purity)

The exact method to prepare an organic solid depends on the actual solid being prepared but, in general, an organic solid can be synthesised and purified using several stages, as shown in Figure 36.

Preparation and separation of the crude product

To prepare an organic solid solutions of the reactants are often added together at room temperature and the product precipitates out. Alternatively the reactants are refluxed together, the anti-bumping granules decanted off and the solid forms on cooling and crystallising.

Suction filtration/filtration under reduced pressure is used to separate the product. It is faster than normal filtration and the solid is left quite dry.

Tip

You also have to know about the preparation of a *pure organic liquid* — this is covered in full detail in required practical 5.

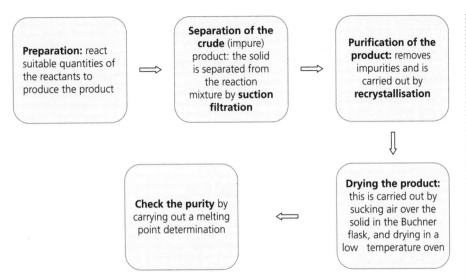

Figure 36 Synthesis of an organic solid

To set up the apparatus for suction filtration, place a circle of filter paper into a Buchner funnel and place in a stopper in a Buchner flask. Connect the Buchner flask to a suction pump and pour the mixture into the funnel. The suction draws the liquid through into the Buchner flask and leaves the crude product in the filter paper (Figure 37).

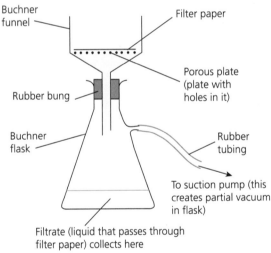

Figure 37 Filtration under reduced pressure

Purification and drying

To obtain a pure solid product, recrystallisation is usually required. A solvent is chosen in which the desired product dissolves readily at higher temperatures but is only slightly soluble at room temperature. The solvent is often water, but it could be another suitable liquid, such as an alcohol. A minimum volume of hot solvent is used to dissolve the solid, making a saturated solution, and any insoluble impurities are then removed by gravity filtration. On cooling, the solubility of the product drops, causing it to recrystallise from solution. The crystals that form are separated

> **Tip**
>
> A *minimum volume of hot solvent* is used to ensure that as much of the solute is obtained as possible.

by filtration under reduced pressure. Impurities remain dissolved in the solution. Slow crystallisation is preferred because fast crystallisation can cause some soluble impurities to become trapped in the crystals.

The method of recrystallisation is as follows:

1 Dissolve the impure crystals in the *minimum volume of hot solvent*. (This ensures that the hot solution is saturated so that crystals form on cooling.)

2 Filter the hot solution by gravity filtration, using a hot funnel and fluted filter paper, to remove any insoluble impurities. (Filtering through a hot filter funnel and using fluted paper prevents precipitation of the solid.)

3 Allow the solution to cool and crystallise. (Reducing the temperature reduces the solubility of the crystals. The impurities will remain in solution.) Sometimes scratching the side of the flask with a glass rod or adding a small seed crystal can aid crystal formation.

4 Filter off the crystals using suction filtration.

5 Wash by pouring over some *ice-cold solvent*, which removes any aqueous impurities. (The solvent is cold to prevent the crystals from dissolving.)

6 Dry by sucking air over the crystals in the Buchner flask and then in a low-temperature oven. Alternative methods of drying include placing in a **desiccator** with a drying agent (Figure 38).

Knowledge check 32

Name the method used to purify some solid paracetamol.

Tip

A clean cork can be used to compress the crystals in the Buchner funnel. This ensures that air passes through the sample of crystals and not just round it, giving better drying.

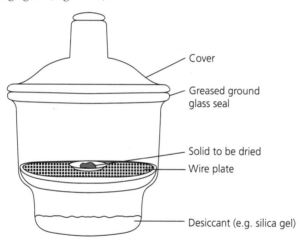

Cover

Greased ground glass seal

Solid to be dried
Wire plate

Desiccant (e.g. silica gel)

Figure 38 A desiccator

Further recrystallisation can be carried out to obtain a purer product.

Checking the purity of an organic solid

The melting point of a substance is not the exact point at which it melts but rather the range of temperatures from when the sample starts to melt until it has completely melted. The greater the range the more impurities are present. A pure substance melts over a narrow range of temperatures while an impure substance melts over a wide range of temperatures and at a lower temperature than the pure substance. The melting points of almost all substances are available in data tables.

To check the purity of a solid, a melting point can be determined using the following method:

Tip

A melting point range of less than 2°C indicates a fairly pure substance.

1 Place some of the solid in a melting point tube *sealed* at one end.

2 Place in melting point apparatus and heat slowly.

3 Record the temperature at which the solid starts to melt and the temperature at which it finishes melting.

4 Repeat and average the temperatures.

5 Compare the melting point with known values in a data book.

Often when preparing a solid compound, the melting point is lower than that given in a data book — this may be due to impurities such as water. It is best to ensure that solids are very dry before carrying out a melting point test.

Percentage yield

Why is the percentage yield less than 100% when preparing an organic solid?

The theoretical reasons are similar as for an organic liquid. The practical reasons why some solid is lost are:

■ Some product is lost in purification steps, for example in recrystallisation some solid may still be dissolved in the solvent.

■ Some product is lost in transferring between vessels — sometimes rinsing is useful to minimise this loss.

Every organic solid will be prepared using a slightly different method. It is important that you understand the general preparation methods listed and apply them to different situations.

Worked example

The following method can be used to prepare aspirin in the laboratory:

■ Place 20.0 g of 2-hydroxybenzoic acid in a pear-shaped flask and add 40 cm³ of ethanoic anhydride (($CH_3CO)_2O$).

■ Safely add 5 cm³ of conc. phosphoric(v) acid and heat under reflux for 30 minutes.

■ Add water to hydrolyse any unreacted ethanoic anhydride to ethanoic acid, and pour the mixture into 400 g of crushed ice in a beaker.

■ The product is removed by filtration under reduced pressure, recrystallised, washed with ice-cold water and dried in a desiccator.

■ The melting point is then determined.

The reaction can be represented as follows:

$$HOOCC_6H_4OH + (CH_3CO)_2O \rightarrow HOOCC_6H_4OCOCH_3 + CH_3COOH$$

a Suggest the role of concentrated phosphoric(v) acid in this preparation.

b Explain how the concentrated phosphoric(v) acid is added safely.

c What is used to heat the mixture under reflux?

d What happens to the ethanoic acid formed by hydrolysis of ethanoic anhydride?

e Assuming a 70% yield, calculate the mass of 2-hydroxybenzoic acid required to form 5.0 g of pure aspirin.

f Why is the mixture poured onto crushed ice?

g Describe the properties required of a suitable solvent for use in recrystallisation.

→

Knowledge check 33

Suggest how you would know that a sample of aspirin was impure from the melting point determination.

Tip

The *identity* of an organic substance can be confirmed by taking a melting point of a pure sample and comparing it with known values in a data book.

Knowledge check 34

Write an equation to define percentage yield.

h State what happens to insoluble and soluble impurities during recrystallisation.

i Samples of 2-hydroxybenzoic acid, pure aspirin, the crude product and the recrystallised product from this experiment were spotted onto a chromatography plate. Describe how the plate is analysed.

Answer

a It is a catalyst.

b It is added slowly, with stirring, to safely distribute any heat produced.

c To safely heat under reflux use a boiling water bath/sand bath/electric heater.

d Ethanoic acid is miscible with water, so it will be in the filtrate.

e moles of aspirin $= \dfrac{5.0}{180.0} = 0.0278$

There is a 70% yield, so:

$\dfrac{0.0278}{70} \times 100 = 0.040$

number of moles 2-hydroxybenzoic acid $= 0.040$

mass of 2-hydroxybenzoic acid $= 0.040 \times 138.0 = 5.52\,g$

f To form crystals.

g A solvent is chosen in which the aspirin is more soluble at higher temperatures and less soluble at lower temperatures.

h Insoluble impurities are filtered out and soluble impurities remain dissolved in the solvent.

i R_f values are taken. The crude product contains a spot that is 2-hydroxybenzoic acid and a spot that is aspirin. The recrystallised product contains aspirin and a less intense spot due to unreacted 2-hydroxybenzoic acid as there is less impurity present. The pure aspirin only contains aspirin.

Required practical 11

Carry out simple test-tube reactions to identify transition metal ions in aqueous solution

Transition metal ions in aqueous solution can be identified in qualitative analysis using sodium hydroxide, ammonia or sodium carbonate solution (Table 11).

Table 11

Metal ion solution	Cu^{2+}	Fe^{2+}	Fe^{3+}
Effect of adding NaOH(aq) until in excess	Blue ppt; insoluble in excess reagent	Green ppt; insoluble in excess reagent	Brown ppt; insoluble in excess reagent
Effect of adding NH_3(aq) until in excess	Blue ppt that dissolves in excess to form a dark blue solution	Green ppt; insoluble in excess reagent	Brown ppt; insoluble in excess reagent
Effect of adding Na_2CO_3(aq)	Green ppt of copper carbonate	Green ppt	Brown ppt; bubbles of gas

Tip

The green precipitate of iron(ii) hydroxide formed is oxidised by air to brown iron(iii) hydroxide.

Aluminium ions in solution can also be identified by adding similar solutions: a white ppt is produced with sodium hydroxide, and dissolves in excess; a white ppt is produced with ammonia solution; and a white ppt with bubbles of gas is produced with sodium carbonate solution. Note that aluminium is not considered as a transition metal.

Knowledge check 35

Describe how sodium carbonate solution can be used to distinguish between a solution of iron(ıı) ions and a solution of iron(ııı) ions.

Worked example

To determine the identity of a chemical salt A, the following tests were carried out and the observations recorded.

Test	Observation
1 Make a solution of A by dissolving in $50\,cm^3$ of deionised water	Blue crystals dissolve to form a blue solution
2 Add three drops of barium nitrate solution to $2\,cm^3$ of a solution of A	White ppt in a blue solution
3 Add three drops of silver nitrate solution to $2\,cm^3$ of a solution of A	No change
4 Add sodium hydroxide solution to $2\,cm^3$ of a solution of A, dropwise with shaking, until in excess	Blue ppt
5 In a fume cupboard, add concentrated ammonia solution dropwise, until present in excess, to $2\,cm^3$ of the solution of A	Blue ppt, which dissolves in excess to produce a deep blue solution
6 Add a few drops of sodium carbonate solution to $2\,cm^3$ of a solution of A	Green ppt

a Use the evidence in the table to suggest the metal ion present in A.

b Suggest the formula of the complex ion formed in test 2. Name the shape of this complex.

c Using the results from tests 3 and 4, determine the anion present in A. State your reasons, giving an ionic equation for any reaction that occurs.

d State the formula of the complex formed in test 5.

e Explain, using two balanced symbol equations, the reaction occurring in test 6.

f What is the name of the green precipitate formed in test 7?

g Suggest and explain what would be observed if salt A was heated gently in a boiling tube.

h Suggest a name for A.

Answer

a Test 1 produces a blue solution, which is the first indication of copper ion. Tests 4, 5 and 6 are consistent with a deduction of Cu^{2+}.

b $[Cu(H_2O)_6]^{2+}$; octahedral

c From test 3, a halide ion is not present. In test 4, insoluble barium sulfate is formed as a white ppt in the blue solution. Solution A contains a sulfate ion.

$$Ba^{2+} + SO_4^{2-} \rightarrow BaSO_4$$

d $Cu(H_2O)_4(OH)_2$

e $[Cu(H_2O)_6]^{2+} + 2OH^- \rightarrow Cu(H_2O)_4(OH)_2$

$Cu(H_2O)_4(OH)_2 + 4NH_3 \rightarrow [Cu(NH_3)_4(H_2O)_2]^{2+} + 2H_2O + 2OH^-$

f Copper carbonate

g The solid is in the form of crystals, so it is hydrated. On heating, the crystals should turn to powder and beads of colourless liquid should form on the test tube.

h Hydrated copper(II) sulfate

Required practical 12

Separation of species by thin-layer chromatography

Chromatography separates the components of a mixture. In thin-layer chromatography the **stationary phase** is a thin layer of either aluminium oxide, silicon(IV) oxide or silica gel, which is supported on a glass plate. The **mobile phase** is a solvent.

The method is as follows:

1 Draw a pencil line 1.5 cm from the bottom of the thin-layer chromatography plate and place two pencil crosses on the line.

2 Place a drop of the purified solid on a watch glass and dissolve in a few drops of solvent such as ethanol. Use a capillary tube to place a spot of the solvent on a pencil cross. Allow the spot to dry and repeat 3–4 times, ensuring that the diameter of the spot is no more than 0.5 cm. This produces a concentrated spot. Repeat this for the pure solid.

3 Place solvent in a beaker/developing tank to a depth of 1 cm.

4 Place the TLC plate in the beaker/tank and cover with a lid.

5 Allow the solvent to run up the chromatography plate for about 30 minutes. When it has almost reached the top of the plate, remove from the beaker and mark the line of the solvent front with a pencil.

6 Place the plate in a fume cupboard until all of the solvent has evaporated and the plate is dry. If the spots are colourless then there are different ways to view them.

- Place the plate under a UV lamp and mark the locations of the substances using a pencil.

- In a fume cupboard place the plate in a beaker containing iodine crystals and cover with a watch glass. The iodine is a locating agent which causes the spots to become brown and visible.

- Spray with ninhydrin developing agent in a fume cupboard.

7 Measure the distance from the pencil line to the spot, and measure the distance from the pencil line to the solvent front. Calculate the R_f value of each substance visible on the plate and compare to data books.

Worked example

a Why is it necessary to wear plastic gloves when holding a TLC plate?

b Why is it necessary to draw a *pencil* base line 1.5 cm from the bottom of the plate?

c How is a very tiny concentrated drop of amino acid solution added to the TLC plate?

d Explain why the developing tank/beaker solvent is at a depth of only 1 cm.

The **stationary phase** in chromatography may be a solid or a liquid held by a solid support.

The **mobile phase** moves through the stationary phase and may be a liquid or a gas.

Tip

Do not look directly into the UV light.

Tip

In a solvent system a substance should always have the same R_f value, hence allowing comparison.

Tip

TLC can be used to monitor the course of a reaction by taking samples from a reaction mixture at regular intervals, spotting them on a TLC plate, and running this alongside controls of the organic reactant and product.

e Explain why the developing tank/beaker is sealed with a lid when the TLC plate is placed in it.

f Explain why the TLC plate is allowed to dry in a fume cupboard.

g The substance used to develop the spots on the chromatogram is ninhydrin. State one safety precaution specific to the use of this substance, apart from the use of goggles.

Answer

a To prevent contamination of the plate by amino acids from the skin, which would interfere with results.

b The pencil line is insoluble and will not move with the solvent or interfere with results.

c Use a capillary tube, and place a tiny drop on the plate. Allow to dry and repeat.

d If the solvent is too deep it will dissolve away the mixture.

e If the tank is open, solvent may evaporate and not advance up the plate. Having a sealed environment also allows the solvent to saturate the atmosphere inside.

f The solvent can be toxic or flammable.

g Avoid breathing the vapour; wear gloves; carry out the procedure in a fume cupboard.

> **Tip**
>
> $$R_f = \frac{\text{distance moved by spot}}{\text{distance moved by solvent}}$$
>
> Always measure to the centre of each spot.
>
> The R_f of spot X in Figure 39, for example, is $\frac{3}{6} = 0.5$.

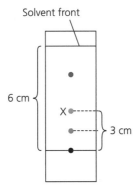

Figure 39 Determing R_f from a chromatogram

Questions & Answers

The AQA A-level exam consists of three papers. All papers will examine practical skills in chemistry. However, paper 3 will have a particular emphasis on practical skills, with 40 marks for practical skills questions, including data analysis.

This Student Guide covers only practical work, mainly through consideration of the 12 required practical activities. Therefore, the questions in this section only reflect the type of question that will be set to test your understanding of experimental methods and are not representative of the papers as a whole.

In the AS exam there are just two papers and questions based on practical work may be set in both papers. **If you are studying for AS then concentrate on questions 1–13 in this section**.

A data sheet is provided with each exam. Copies can be downloaded from the AQA website.

Questions on all papers are a mixture of multiple-choice, short-answer and long-answer structured questions.

You should pay particular attention to diagrams, drawing graphs and making calculations. Many students lose marks by failing to label diagrams properly, by not giving essential data on graphs and, in calculations, by not showing all the working or by omitting units.

In this section, comments preceded by ⓔ indicate what to watch out for when answering the question. Comments on student answers are preceded by the icon ⓔ.

Ticks (✓) are included in the answers to show where a mark has been awarded.

Question 1

A solution of a metal hydroxide XOH was made up by dissolving 3.92 g of solid in 250 cm^3 of water. 25.0 cm^3 of XOH solution was placed in a conical flask with a few drops of bromothymol blue indicator. The indicator is yellow in acid and blue in alkali. The conical flask was placed on a white tile and titrated with a standard solution of hydrochloric acid of concentration 0.500 mol dm^{-3}, in order to find the concentration of XOH.

The balanced symbol equation for the reaction is:

$$XOH + HCl \rightarrow XCl + H_2O$$

(a) Describe how you would safely and accurately measure out and place 25.0 cm^3 of the metal hydroxide solution in the conical flask, naming all the apparatus you would use.

(3 marks)

Student answer

(a) Rinse a pipette with the XOH solution ✓. Using a pipette ✓ and pipette filler ✓ transfer 25.0 cm^3 of this solution into a conical flask.

ⓔ **3/3 marks awarded** To measure out $25.0\,cm^3$ accurately a pipette is used, and a safety pipette filler for safety. For accuracy, rinse the pipette with the XOH.

(b) Why is a white tile used in this practical technique? (1 mark)

(b) To aid detection of the colour change ✓.

(c) State the colour change of the bromothymol blue indicator at the end point. (1 mark)

(c) Blue to yellow ✓

ⓔ **1/1 mark awarded** XOH is an alkali and is in the conical flask, so the colour change will be from blue to yellow.

(d) Describe how the end point was accurately determined. (1 mark)

(d) Add acid dropwise ✓, with swirling until the colour changes.

ⓔ **1/1 mark awarded** To determine the end point accurately, after a trial titration has been carried out, run the solution in until near the trial value, and then add in drops, with swirling.

(e) Why is the conical flask swirled during a titration? (1 mark)

(e) To mix the solution and ensure that the solutions have reacted ✓.

(f) If the burette was not filled correctly, and the gap between the tap and the tip of the burette still contained air, suggest the effect this would have on the titre. (1 mark)

(f) The titre would be greater ✓.

ⓔ **1/1 mark awarded** When the burette is opened, the tap is filled before any liquid is delivered into the flask, hence the measured volume would be greater.

The results obtained in this experiment are shown below.

	Trial	1	2	3
Final burette reading/cm^3	14.40	28.25	42.35	14.30
Initial burette reading/cm^3	0.00	14.30	28.25	0.00
Volume of HCl used/cm^3	14.40	13.95	14.10	14.30

(g) Calculate the concentration of the XOH in $mol\,dm^{-3}$. Give your answer to two significant figures. (3 marks)

(g) average titre $= \dfrac{13.95 + 14.10}{2} = 14.0$ ✓

moles HCl $= \dfrac{14.0 \times 0.5}{1000} = 0.007$

ratio $1:1$; moles XOH $= 0.007$

$\dfrac{0.007}{0.025} = 0.28\,mol\,dm^{-3}$ ✓✓

@ **3/3 marks awarded** First calculate the average titre using concordant values
— 13.95 and 14.10. Then calculate the amount in moles of acid and, by ratio, the
moles of XOH.

(h) Calculate the relative formula mass (M_r) of XOH and identify the element X. (3 marks)

(h) From the initial weighings:

3.92 g in 250 cm^3

$3.92 \times 4 = 15.68$ g dm^{-3}

$\dfrac{15.68 \text{ g dm}^{-3}}{0.28 \text{ mol dm}^{-3}} = 56$

XOH = 56 ✓✓

X = 56 – 17 = 39. Element X is K ✓

(i) A student obtained a result for X and her teacher told her it was lower than
expected. Suggest a reason for this. (1 mark)

(i) The solid XOH may absorb water from the air/the XOH may react with
carbon dioxide in the air ✓.

Question 2

**Which of the following pieces of apparatus has the lowest percentage error in the
measurement shown?** (1 mark)

A Volume of 25.0 cm^3 measured with a burette with an error of +0.1 cm^3.

B Volume of 20 cm^3 measured with a measuring cylinder with an error of +1 cm^3.

C Mass of 0.320 g measured with a balance with an error of +0.001 g.

D Temperature change of 83.2°C measured with a thermometer with an error
of +0.1°C.

@ Use the following equation to calculate the percentage uncertainty:

$$\text{percentage uncertainty} = \frac{\text{uncertainty}}{\text{quantity measured}} \times 100$$

Student answer

The correct answer is D ✓.

@ **1/1 mark awarded** $\mathbf{A} = \dfrac{0.1}{25} \times 100 = 0.4\%$

$\mathbf{B} = \dfrac{1}{20} \times 100 = 5\%$

$\mathbf{C} = \dfrac{0.001}{0.320} \times 100 = 0.3\%$

$\mathbf{D} = \dfrac{0.1}{83.2} \times 100 = 0.1\%$

Question 3

Commercial vinegar contains ethanoic acid. $10.0\,cm^3$ of commercial vinegar was diluted to $1\,dm^3$ with water in a volumetric flask, before being used in a restaurant. $25.0\,cm^3$ of the diluted solution was titrated against $0.100\,mol\,dm^{-3}$ sodium hydroxide using phenolphthalein indicator.

(a) Name the piece of apparatus used to measure $25.0\,cm^3$ of diluted solution into the conical flask. (1 mark)

> **Student answer**
>
> **(a)** Pipette ✓

(b) State the colour change of the indicator at the end-point. (1 mark)

The results for the titration are shown in the table.

	Rough titration	1	2	3
Initial burette reading/cm³	0.00	0.00	0.00	17.45
Final burette reading/cm³	18.10	17.55	17.45	35.15
Titre/cm³	18.10	17.55	17.45	17.70

> **(b)** Colourless to pink ✓

ⓔ **1/1 mark awarded** Phenolphthalein is colourless in acid. Do not use the word 'clear' for colourless.

(c) Calculate the mean titre and justify your choice of titres. (2 marks)

ⓔ Use concordant titres to calculate the mean titre.

> **(c)** Using concordant titres — titrations 1 and 2 ✓:
>
> mean titre = $\dfrac{17.55 + 17.45}{2}$ = $17.50\,cm^3$ ✓

(d) Write the equation for the reaction of ethanoic acid with sodium hydroxide. (1 mark)

> **(d)** $CH_3COOH + NaOH \rightarrow CH_3COONa + H_2O$ ✓

(e) Calculate the concentration of the diluted vinegar solution. (2 marks)

ⓔ In this calculation you first need to work out the number of moles of sodium hydroxide and then, by ratio, the moles of ethanoic acid and the concentration of vinegar. The answer can be given to three significant figures because all the data are to three significant figures.

> **(e)** moles sodium hydroxide = $\dfrac{17.50 \times 0.10}{1000}$ = 0.00175
>
> ratio 1 : 1 moles
>
> moles ethanoic acid = 0.000175 ✓
>
> moles = 0.00175 = $\dfrac{25.0 \times conc.}{1000}$
>
> conc. = $0.070\,mol\,dm^{-3}$ ✓

(f) Calculate the concentration of the commercial vinegar solution in $g\,dm^{-3}$. (2 marks)

ⓔ Remember that $1\,dm^3 = 1000\,cm^3$. The dilution factor is $10.0 \to 1000\,cm^3$, which is ×100. To find the undiluted commercial vinegar concentration, multiply by 100 and change to $g\,dm^{-3}$ by multiplying by M_r (60).

> **(f)** $g\,dm^{-3} = 0.070 \times 100 \times 60 = 420\,g\,dm^{-3}$ ✓✓

(g) The uncertainty in the mean titre is 0.15. Calculate the percentage uncertainty. (1 mark)

ⓔ percentage uncertainty $= \dfrac{\text{uncertainty}}{\text{quantity measured}} \times 100$

> **(g)** percentage uncertainty $= \dfrac{0.15}{17.50} \times 100 = 0.86\%$ ✓

(h) Suggest how, using the same solution of ethanoic acid, the experiment could be improved to reduce the percentage uncertainty. (2 marks)

> **(h)** Use a lower concentration of NaOH ✓; so that a larger titre is required ✓.

ⓔ **2/2 marks awarded** In calculating the percentage uncertainty, the uncertainty value is divided by the mean titre, so if the titre is larger, then the percentage uncertainty will be smaller. To get a larger titre a lower concentration of sodium hydroxide must be used.

Question 4

(a) 2.68 g of hydrated sodium carbonate ($Na_2CO_3.xH_2O$) was dissolved and made up to a solution in $250.0\,cm^3$ of deionised water. $25.0\,cm^3$ of this solution was titrated with $0.200\,mol\,dm^{-3}$ hydrochloric acid and the average titre was $10.00\,cm^3$. The equation for the reaction is:

$$2HCl + Na_2CO_3 \to 2NaCl + H_2O + CO_2$$

Determine the value of x in $Na_2CO_3.xH_2O$. (6 marks)

ⓔ In this example the amount in moles of water of crystallisation in hydrated sodium carbonate is determined by making up a solution and titrating samples with hydrochloric acid.

Student answer

(a) moles HCl $= \dfrac{\text{volume } (cm^3) \times \text{conc.}}{1000} = \dfrac{10.00 \times 0.200}{1000} = 0.002$

ratio $= 2HCl : 1Na_2CO_3$

$0.002 : ?$

There are half as many moles of sodium carbonate as hydrochloric acid, so divide the moles of HCl by two:

moles $Na_2CO_3 = \dfrac{0.002}{2} = 0.001$ ✓

0.001 moles Na_2CO_3 in $25.0\,cm^3$

$0.001 \times 10 = 0.01$ moles Na_2CO_3 in $250.0\,cm^3$ ✓

$$moles = \frac{mass}{M_r}$$

$M_r\ Na_2CO_3 = 106$

$$0.01 = \frac{mass}{106}$$

$0.01 \times 106 = 1.06\,g$ of Na_2CO_3 ✓

From the question, the mass of hydrated sodium carbonate ($Na_2CO_3.xH_2O$) in $250.0\,cm^3 = 2.68\,g$.

mass of anhydrous sodium carbonate $Na_2CO_3 = 1.06\,g$

mass of water $= 2.68 - 1.06 = 1.62\,g$ ✓

$$moles\ H_2O = \frac{mass}{M_r} = \frac{1.62}{18} = 0.09$$ ✓

ratio of moles Na_2CO_3 : moles H_2O

$$= 0.01 : 0.09$$
$$= 1 : 9$$

The value of x for this sample of hydrated sodium carbonate is 9 ($Na_2CO_3.9H_2O$) ✓.

ⓔ **6/6 marks awarded** Another method of carrying out the calculation above is to find the concentration of sodium carbonate in $mol\,dm^{-3}$ from the titration results, and then find the concentration in $g\,dm^{-3}$ from the original weighings and use the equation $mol\,dm^{-3} = g\,dm^{-3}/M_r$ to find the M_r. Then subtract the $M_r\ Na_2CO_3$ and divide by 18 ($M_r\ H_2O$) to find x.

(b) Describe a different method that could be used to determine the value of x in $Na_2CO_3.xH_2O$. (5 marks)

(b) 1 Weigh a crucible/evaporating basin ✓.

2 Add hydrated solid and weigh the crucible/evaporation basin with the solid in it ✓.

3 Heat to constant mass — this means heat for a few minutes, cool then weigh and repeat this until the mass does not change ✓.

4 Subtract the mass of the evaporating basin to find the mass of anhydrous sodium carbonate, and then find the moles ✓.

5 Subtract the mass of the evaporating basin + anhydrous sodium carbonate from the mass of the evaporating basin + hydrated sodium carbonate to find the mass of water, and then find the moles ✓.

6 Find the ratio of moles of anhydrous sodium carbonate to moles of water.

Question 5

(a) Describe how a value for the enthalpy change of combustion of ethanol can be determined experimentally. (6 marks)

ⓔ Ethanol is a liquid and the method will involve placing ethanol in a spirit burner and heating a known mass of water. You need to give clear and logical steps in your answer and ensure you describe how the results are used to determine the enthalpy change.

Student answer

(a) Measure the initial temperature of the water ✓.

Use a known volume of water in a calorimeter ✓.

Burn a measured mass of fuel in a spirit burner ✓.

Measure the final temperature of the water using a thermometer and calculate temperature change ✓.

Calculate the number of moles of fuel used using mole = mass/M_r ✓.

Use $q = mc\Delta T$ and scale up for 1 mole of fuel ✓.

(b) When the enthalpy of combustion of pentan-1-ol is determined its value is much larger. Suggest a reason for this. (1 mark)

(b) There are more bonds to be broken and made ✓.

ⓔ **1/1 mark awarded** Each successive member of the homologous series adds an extra –CH_2 group hence pentan-1-ol has more bonds to be broken and made during combustion and so there is a greater value.

(c) The value for the enthalpy change of combustion determined experimentally is frequently much less exothermic than the value found in data books. One reason for this is heat loss. Suggest *two* other reasons. (2 marks)

(c) Incomplete combustion ✓; evaporation of fuel from wick of the burner ✓; not carried out under standard conditions ✓. *(Any 2)*

Question 6

When an ionic solid dissolves in water there may be a temperature change. A student placed $100\,cm^3$ of water in a polystyrene beaker and then recorded the temperature. He powdered some hydrated copper(II) sulfate, $CuSO_4.5H_2O$, and then dissolved approximately 2.10 g of it in the water and recorded the final temperature.

      mass of hydrated copper(II) sulfate added = 2.07 g

      initial temperature = 18.0°C

      final temperature = 17.8°C

      specific heat capacity of water = $4.18\,J\,°C^{-1}\,g^{-1}$

(a) Why, and how, is the hydrated copper(II) sulfate powdered?　　(2 marks)

> **Student answer**
>
> **(a)** Dissolves faster ✓; using a mortar and pestle ✓.

ⓔ **2/2 marks awarded** To grind up solids a mortar and pestle is used. A powder has a greater surface area and reacts or dissolves faster.

(b) Describe, stating the equipment used, how the student could accurately determine the mass of hydrated copper(II) sulfate added.　　(3 marks)

> **(b)** Use a top-pan balance reading to two decimal places ✓. Weigh the solid in a weighing boat, add the solid to water and reweigh the weighing boat ✓. Subtract the values ✓.

ⓔ **3/3 marks awarded** The recorded value is to two decimal places, so a top-pan balance that reads to two decimal places must be used. When using a balance, it is best to measure the solid in a weighing boat, add to the reaction, and then reweigh the weighing boat and subtract the values.

(c) Calculate the enthalpy change on dissolving the hydrated copper(II) sulfate in the water.　　(1 mark)

ⓔ Note that when working out the enthalpy change it is always the mass of water which is used: $100\,cm^3$ of water has a mass of $100\,g$ because the density of water is $1\,g\,cm^{-3}$. The temperature *change* can be used in °C; there is no need to convert to kelvin. The units of enthalpy change are joules.

> **(c)** energy absorbed by water = $q = mc\Delta T = 100 \times 4.18 \times 0.20 = 84.0\,J$ ✓

(d) Calculate the enthalpy change, in $kJ\,mol^{-1}$, on dissolving 1 mol of hydrated copper(II) sulfate in water.　　(3 marks)

ⓔ When calculating the M_r of hydrated copper(II) sulfate remember to include the five moles of water of crystallisation.

> **(d)** $M_r = 63.5 + 32.1 + (4 \times 16.0) + (5 \times 18) = 249.6$ ✓
>
> moles of hydrated copper(II) sulfate used = $2.07/249.6 = 0.00829$ ✓
>
> $0.00829\,mol$ produces $84.0\,J$
>
> 1 mol produces $84/0.00829 = 10\,132.69\,J\,mol^{-1} = 10.133\,kJ\,mol^{-1}$ ✓
>
> $\Delta H = -10.1\,kJ\,mol^{-1}$

(e) The actual value is $11.7\,kJ\,mol^{-1}$. State *one* source of error in the student's experiment and suggest how it could be reduced.　　(2 marks)

> **(e)** There are energy transfers to and from the surroundings ✓; use a lid/use more insulation ✓.

Question 7

50.0 g of water was measured out and placed in a plastic beaker at 20.0°C. A 5.00 g sample of ammonium chloride was added and the mixture was stirred. As the ammonium chloride dissolved, the temperature of the solution decreased.

Describe the steps you would take to determine an accurate minimum temperature that is not influenced by heat from the surroundings.

(4 marks)

> **Student answer**
>
> Start a clock when ammonium chloride is added to water ✓.
>
> Record the temperature every minute for about 5 minutes ✓.
>
> Plot a graph of temperature against time ✓.
>
> Extrapolate the graph back to the time of mixing ($t = 0$) and determine the temperature ✓.

ℯ 4/4 marks awarded It is best to record the temperature with time and extrapolate the graph to get the minimum temperature.

Question 8

(a) To measure the rate of a reaction for a precipitation reaction, 20 cm³ of solution A was added to a conical flask placed on a cross, and 20 cm³ of solution B added and a timer started. The timer was stopped when it was no longer possible to see the cross on the paper. The experiment was repeated.

What is likely to decrease the accuracy of the experiment?

(1 mark)

- **A** Rinsing the flask with the solution A before each new experiment.
- **B** Stirring the solution throughout each experiment.
- **C** Using the same cross on paper for each experiment.
- **D** Using different measuring cylinders to measure the volumes of each solution.

> **Student answer**
>
> **(a)** The answer is A — rinsing the flask with solution A before each new experiment ✓.

ℯ 1/1 mark awarded In this practical there should be controlled variables. The same cross should be used, so it is of the same darkness in all experiments, and if the reactants are stirred they should be stirred in all experiments. Using different measuring cylinders increases accuracy as it prevents precipitation occurring in the cylinder if some drops of the previous solution are left in the cylinder.

(b) Which statement is correct about the time taken for the cross to disappear if the experiment is repeated using a larger conical flask?

(1 mark)

- **A** The time taken will not be affected by using the larger conical flask.
- **B** The time taken will be decreased by using the larger conical flask.
- **C** The time taken will be increased by using the larger conical flask.

D It is impossible to predict how the time taken will be affected by using the larger conical flask.

> **(b)** The answer is C — the time taken will be increased by using the larger conical flask ✓.

ⓔ **1/1 mark awarded** Using a larger flask means that the precipitation will not form as thick a covering and so it will take longer for the cross to be obscured.

Question 9

Some hydrochloric acid was added to some sodium carbonate solution in one test tube and to some silver nitrate in another test tube. Which letter gives the correct observations?

(1 mark)

	A	B	C	D
Sodium carbonate solution	No change	No change	Effervescence	Effervescence
Silver nitrate solution	Precipitate	No change	No change	Precipitate

> **Student answer**
>
> The answer is D ✓.

ⓔ **1/1 mark awarded** When an acid is added to a carbonate effervescence results. The anion in hydrochloric acid is chloride and it will form a precipitate with silver nitrate solution.

Question 10

Which of the following is a correct statement about the test for bromide ions in solution?

(1 mark)

A Add acidified $AgNO_3(aq)$; white precipitate formed, soluble in dilute ammonia solution.

B Add acidified $AgNO_3(aq)$; cream precipitate formed, soluble in dilute ammonia solution.

C Add acidified $AgNO_3(aq)$; white precipitate formed, soluble in concentrated ammonia solution.

D Add acidified $AgNO_3(aq)$; cream precipitate formed, soluble in concentrated ammonia solution.

> **Student answer**
>
> The answer is D ✓.

ⓔ **1/1 mark awarded** Bromide ions form a cream precipitate with silver nitrate solution. The cream precipitate is not soluble in dilute ammonia but is soluble in concentrated ammonia solution.

Question 11

(a) A solid compound was dissolved in deionised water. Dilute nitric acid was added, followed by silver nitrate solution. A white precipitate was observed, which dissolved when ammonia solution was added.

    (i) Name the ion that has been identified during this test.   (1 mark)

    (ii) Write the simplest ionic equation for the formation of the white precipitate.   (1 mark)

    (iii) Explain why dilute nitric acid is added to the solution before silver nitrate solution.   (2 marks)

ⓔ Silver nitrate solution is the key to this question as it is used to test for halide (chloride, bromide and iodide) ions. Chloride ions give a white precipitate. Be careful that you do not write down chlorine ions — this is incorrect. The dilute nitric acid is added to remove any carbonate ions, which would give a white precipitate of silver carbonate, Ag_2CO_3. This would be a false positive test for chloride ions. Learn the observations carefully in terms of the colour of the precipitates and whether they redissolve in ammonia solution.

Student answer

(a) (i) Chloride ✓

    (ii) $Ag^+ + Cl^- \rightarrow AgCl$ ✓

    (iii) Nitric acid reacts with/removes carbonate/hydroxide ions ✓. Carbonate ions would give a white precipitate/false test for chloride ions ✓.

(b) It is possible to distinguish between $BaCl_2(aq)$ and $MgCl_2(aq)$ by adding an aqueous reagent to a sample. Give a suitable aqueous reagent that could be added separately to each compound. Describe what you would observe in each case.   (3 marks)

ⓔ These are both group 2 compounds. You need to think about identification tests and about the solubility of group 2 compounds.

(b) Add some aqueous sodium sulfate or sulfuric acid (any named soluble sulfate) ✓. A white precipitate forms for barium chloride solution ✓. Whereas the solution remains colourless/no reaction for magnesium chloride solution ✓.

Or

Add sodium or potassium hydroxide solution ✓. The solution remains colourless/no reaction for barium chloride solution ✓. A white precipitate occurs for magnesium chloride solution ✓.

ⓔ 3/3 marks awarded Remember that magnesium hydroxide is insoluble but magnesium sulfate is soluble, and barium hydroxide is soluble but barium sulfate is insoluble. Adding any soluble sulfate — for example sodium sulfate solution or sulfuric acid — to the compounds will form magnesium sulfate, which is soluble and produces a colourless solution, and barium sulfate, which is insoluble and produces a white precipitate. Note that you must give the full name of the reagent — it is not sufficient to write 'add a soluble sulfate'. Alternatively, sodium or potassium hydroxide solution could be added. Magnesium hydroxide is formed as an insoluble white precipitate; barium hydroxide is soluble and a colourless solution remains.

Question 12

The following method can be used to prepare 1-bromobutane:
- Place 30 cm^3 of water, 40 g of powdered sodium bromide and 21.8 g of butan-1-ol in a 250 cm^3 round-bottomed flask.
- Add 25 cm^3 of concentrated sulfuric acid dropwise, cooling the flask in an ice bath.
- When the addition is complete, reflux for 45 minutes and then distil off the crude
  1-bromobutane.
- Wash the distillate with concentrated sulfuric acid and then with sodium carbonate solution.
- Remove the 1-bromobutane layer and add a spatula of anhydrous sodium sulfate;
  swirl and filter.
- Distil and collect the 1-bromobutane at 99–102°C.

(a) Write the equation for the formation of 1-bromobutane from butan-1-ol and hydrogen bromide.   (1 mark)

ⓔ This reaction is not specifically part of the specification, so simply use the information in the question. You know the formula of butan-1-ol and of 1-bromobutane, so put these into an equation.

> **Student answer**
>
> **(a)** $CH_3CH_2CH_2CH_2OH + HBr \rightarrow CH_3CH_2CH_2CH_2Br + H_2O$ ✓

(b) What is reflux?   (1 mark)

> **(b)** Repeated boiling and condensing of a reaction mixture to ensure that reaction takes place without the contents of the flask boiling dry ✓.

(c) State three changes made to the apparatus set-up to change from reflux to distillation.   (3 marks)

> **(c)** Add a still head ✓; add a thermometer ✓; condenser moved from upright to sideways position ✓; receiver on end of condenser ✓. *(Any three)*

ⓔ 3/3 marks awarded Reflux means that the condenser is in the vertical position. For distillation the condenser is sideways, and a still head connector and receiver need to be used. You should be able to draw this apparatus, and also to describe it.

(d) Suggest why sodium carbonate solution is added to wash the distillate, rather than sodium hydroxide solution. (2 marks)

> **(d)** Sodium carbonate solution is a weaker alkali than sodium hydroxide solution ✓. Sodium hydroxide solution would hydrolyse the halogenoalkane ✓.

ⓔ 2/2 marks awarded First think of the difference between sodium carbonate solution and sodium hydroxide solution — the sodium carbonate is a weaker alkali. Halogenoalkanes can be hydrolysed by strong alkali, hence it is better to add sodium carbonate solution, which neutralises the acid impurities but does not hydrolyse the product.

(e) Describe practically how the distillate is washed with aqueous sodium carbonate. (4 marks)

ⓔ Washing of an organic liquid to purify it is carried out in a separating funnel. Ensure that you mention safety precautions, such as release of pressure. This method must be learned in detail.

> **(e)** Add the distillate to a separating funnel with a portion of aqueous sodium carbonate ✓.
>
> Stopper and shake, inverting the funnel and opening the tap to release the pressure ✓.
>
> Allow the separating funnel to stand until the layers settle and separate. Remove the stopper and open the tap, to run off the bottom layer into a beaker. Run off the second layer into a separate beaker and discard the aqueous layer ✓.
>
> Place the organic layer back into the separating funnel and repeat the process, using another portion of the aqueous solution ✓.

(f) Without using densities, how would you determine which layer is the aqueous layer during washing? (1 mark)

ⓔ Remember that an aqueous solution is a solution in water.

> **(f)** Add water to the separating funnel. The layer that increases in size is the aqueous layer ✓.

Question 13

Ethanol is oxidised to ethanoic acid by heating under reflux using acidified potassium dichromate(VI) solution. What is the reason for heating under reflux? (1 mark)

A To allow efficient mixing of the solutions.

B To boil the mixture at a higher temperature.

C To ensure even heating.

D To prevent any substances escaping.

Student answer

The answer is D ✓.

ⓔ **1/1 mark awarded** Reflux means the water-cooled condenser is in the vertical position. Mixing of the solutions could be achieved by shaking or agitation — the vertical condenser does not help with this. It does not allow the mixture to boil at a higher temperature — impurities added would do this, and anti-bumping granules are needed to ensure even heating. The reason for reflux is to ensure that any volatile liquids do not escape, hence answer D is correct.

Question 14

(a) A student carried out an experiment to determine which of a chloroalkane, bromoalkane or iodoalkane hydrolysed fastest when heated with silver nitrate in ethanol. Describe an experimental method he could use. Ensure that it is a fair test.

(6 marks)

Student answer

(a) *Any two fair-test marks:*

■ Use haloalkanes with the same chain length ✓.

■ Equal amounts of ethanol, silver nitrate mixture in each test tube and equal volumes of haloalkane ✓.

■ All tubes at the same temperature ✓.

Any four methods marks:

■ Place (equal volumes of) each haloalkane in separate test tubes and place in a water bath at 50°C ✓.

■ Heat a test tube of a mixture of ethanol and aqueous silver nitrate in the same water bath ✓.

■ When all test tubes have reached the same temperature add (equal volumes of) the mixture to each haloalkane and start the clock ✓.

■ Time how long it takes for each precipitate to form ✓.

ⓔ **6/6 marks awarded** In this reaction the haloalkane hydrolyses and halide ions are produced. Hence, a precipitate of silver halide gradually appears when the halide ions react with the silver nitrate. The rate of precipitation is a measure of the rate of hydrolysis. You may name the haloalkanes used, e.g. chlorobutane, bromobutane and iodobutane, or refer to the fact that they have the same chain length, for a fair test. Note that a water bath must be used to heat because of the flammable nature of ethanol.

(b) State and explain the results, in terms of ease of hydrolysis of the haloalkanes, that you would expect in this experiment. (2 marks)

> **(b)** The iodoalkane forms a precipitate first, then the bromoalkane, then the chloroalkane ✓. The C–I bond energy is smaller than the C–Br and C–Cl bond energies ✓.

(c) Describe how you would identify the chloroalkane, bromoalkane and iodoalkane in this experiment. (1 mark)

> **(c)** White ppt: chloroalkane
>
> Cream ppt: bromoalkane
>
> Yellow ppt: iodoalkane ✓

(d) The densities and the boiling points of haloethanes are listed in the table below.

Haloethane	Density/g cm^{-3}	Boiling point/°C
Chloroethane	0.898	12
Bromoethane	1.461	38
Iodoethane	1.936	72

   (i) Suggest why there is an increase in density from bromoethane to iodoethane. (1 mark)

   (ii) Suggest why there is an increase in boiling point from chloroethane to iodoethane. (2 marks)

   (iii) If all three haloethanes were present in a container at room temperature, suggest how you would separate and obtain each haloethane. (2 marks)

ⓔ Look carefully at the data to help you interpret this. At room temperature the chloroethane is a gas and the other two are liquids.

> **(d) (i)** Increase in mass of halogen atom ✓

ⓔ **1/1 mark awarded** The haloethanes have different atoms and hence different masses. The increase in density is because the halogen atom increases in mass.

> **(ii)** More electrons/greater M_r in iodoethane ✓
> Stronger van der Waals forces between molecules ✓

ⓔ **2/2 marks awarded** Boiling point can be explained by looking at the intermolecular forces. Due to the larger relative molecular mass in iodoethane, there are stronger van der Waals forces and more energy is needed to break these.

> **(iii)** Chloroethane is a gas and must be removed/cool to liquify ✓. Fractional distillation — collect at boiling points ✓.

Question 15

Some tests were carried out on an organic liquid, B, which has the *empirical* formula C_2H_4O. The results are shown in the table below.

Test	Observation
1 Add $2\,cm^3$ of deionised water to $2\,cm^3$ of B in a test tube.	One layer forms
2 Add 10 drops of B to $2\,cm^3$ of acidified potassium dichromate solution in a test tube. Place the test tube in a hot water bath.	Solution remains orange
3 Place $5\,cm^3$ of B in a boiling tube. Add $5\,cm^3$ of ethanol, and then $1\,cm^3$ of concentrated sulfuric acid. Heat the boiling tube in a water bath. Cautiously smell the contents of the boiling tube.	Sweet smell
4 Add a spatula measure of sodium carbonate to $2\,cm^3$ of B in a test tube.	Effervescence

(a) What can be deduced from test 1? (1 mark)

> **Student answer**
>
> **(a)** B is miscible with water/–OH group present/B can hydrogen bond with water ✓.

ⓔ 1/1 mark awarded If an organic substance forms one layer with water, it means it is miscible. This is because of hydrogen bonds forming between the organic compound and the water, due to the organic compound possibly containing an –OH group.

(b) What can be deduced from test 2? (1 mark)

> **(b)** B is not a primary or secondary alcohol, or aldehyde ✓.

ⓔ 1/1 mark awarded Acidified potassium dichromate changes colour from orange to green when warmed with an aldehyde, primary alcohol or secondary alcohol. There is no colour change and so B is not one of these, but may be a tertiary alcohol or a carboxylic acid.

(c) What can be deduced from test 3? (1 mark)

> **(c)** B may be a carboxylic acid/an ester has formed ✓.

ⓔ 1/1 mark awarded The sweet-smelling substance is the ester formed when the ethanol reacts with B in the presence of a catalyst of concentrated sulfuric acid. It is likely that B is a carboxylic acid.

(d) What can be deduced from test 4? (1 mark)

> **(d)** B is a carboxylic acid. ✓

(e) Describe a test that could be used to test for the gas produced in test 4. (2 marks)

> **(e)** Bubble the gas into limewater ✓. It changes from colourless to cloudy ✓.

(f) The mass spectrum of B is shown below.

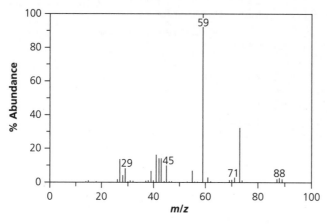

Mass spectrum of B

Draw a displayed structure for B. (1 mark)

ⓔ B is a carboxylic acid. It has the empirical formula C_2H_4O. The mass spectrum shows that it has a relative molecular mass of 88, so its molecular formula is $C_4H_8O_2$.

> **(f)**
>
> H—C—C—C—C—O—H   *or*   H—C—C—C—O—H
>
> ✓

(g) Identify the species responsible for the base peak in the spectrum above. (1 mark)

> **(g)** $CH_2COOH^+/CH_2COOH^+/C_2H_3O_2^+$ ✓

Question 16

Describe chemical tests that you could carry out in test tubes to distinguish between compounds X, Y and Z. Include appropriate reagents and any relevant observations. Also include equations showing structures for the organic compounds involved. (6 marks)

X   Y   Z

Student answer

Warm all solutions with Tollens' reagent and a silver mirror is formed with Y, or warm with Fehling's solution and a red ppt is formed with Y ✓.

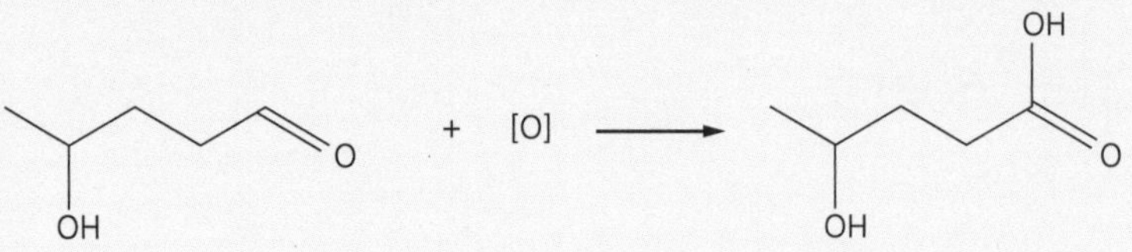

✓

Then warm fresh samples of X and Z with acidified potassium dichromate(VI) ✓. The orange solution changes to green in X, which is a primary alcohol ✓.

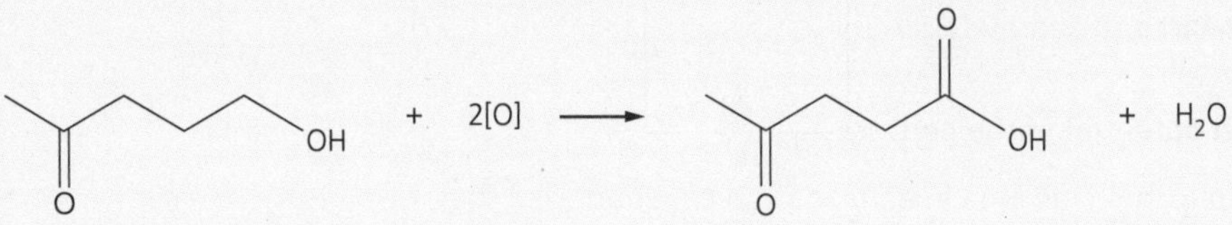

✓

The orange solution stays orange in Z (tertiary alcohol) ✓.

ⓔ **6/6 marks awarded** Compound Y has a carbonyl group at the end of the chain and is an aldehyde, which will react with Tollens' reagent or Fehling's solution. X and Z both have alcohol groups, but X is a primary alcohol, so it will be oxidised by acidified potassium dichromate(VI) and Z is a tertiary alcohol and cannot be oxidised. The order of reactions can be different.

Question 17

In an experiment to investigate cells, a student set up a cell between a silver half-cell and a zinc half-cell under standard conditions. The standard redox potential for each half-cell is given below:

$Ag^+(aq) + e^- \rightarrow Ag(s)$ +0.80 V

$Zn^{2+}(aq) + 2e^- \rightarrow Zn(s)$ −0.76 V

(a) Outline the experimental setup that could be used in the laboratory to measure the standard cell potential of the cell. In your answer you should include a diagram of the set-up and the standard conditions required.   (4 marks)

ⓔ To experimentally determine a standard cell potential, each half-cell must consist of the metal dipping into a $1\,mol\,dm^{-3}$ solution of the metal ions. A high-resistance voltmeter is used and a salt bridge.

Student answer

(a)

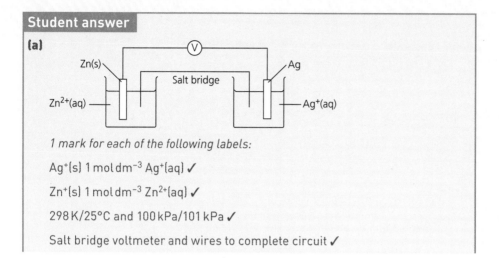

1 mark for each of the following labels:

$Ag^+(s)$ $1\,mol\,dm^{-3}$ $Ag^+(aq)$ ✓

$Zn^+(s)$ $1\,mol\,dm^{-3}$ $Zn^{2+}(aq)$ ✓

$298\,K/25°C$ and $100\,kPa/101\,kPa$ ✓

Salt bridge voltmeter and wires to complete circuit ✓

(b) Calculate the standard cell potential of the cell. (1 mark)

e The zinc loses electrons and forms zinc ions in an oxidation reaction. Use the equation: $E_{cell} = E_{reduction\ reaction} - E_{oxidation\ reaction}$

(b) $Zn^{2+}(aq) + 2e^- \rightleftharpoons Zn(s)$ $E^\ominus = -0.76\,V$

$Ag^+(aq) + e^- \rightleftharpoons Ag(s)$ $E^\ominus = +0.80\,V$

$E_{cell} = E_{reduction\ reaction} - E_{oxidation\ reaction}$

$E^\ominus_{cell} = +0.80 - (-0.76) = +1.56\,V$ ✓

(c) Indicate the direction of flow of electrons in the external circuit. (1 mark)

(c) The electrons move from the zinc half-cell through the external circuit to the silver half-cell ✓.

(d) Why is a salt bridge made of filter paper saturated with potassium chloride solution not effective as a salt bridge in this reaction? (1 mark)

e Think about the ions that are present in the aqueous solutions in the beaker — silver ions and zinc ions.

(d) The chloride ions would react with the silver ions, precipitating silver chloride ✓.

e **1/1 mark awarded** If potassium chloride was used as salt bridge then the silver ions may react with the chloride ions, forming insoluble silver chloride.

(e) Suggest a change to one half-cell that would increase the overall cell potential. (1 mark)

(e) Decreasing the concentration of zinc ions/increasing the concentration of silver ions ✓.

e **1/1 mark awarded** By decreasing the concentration of the zinc ions, the position of the zinc equilibrium will shift to the left, making the electrode potential more negative. This will increase the cell potential. Equally, increasing the concentration of the silver ions will shift the position of the silver equilibrium to the right, making the electrode potential more positive, and also increasing the magnitude of the cell potential.

Question 18

Bromine reacts with aqueous methanoic acid in a reaction that is catalysed by hydrogen ions:

$$Br_2(aq) + HCOOH(aq) \xrightarrow{H^+(aq)} 2Br^-(aq) + 2H^+(aq) + CO_2(g)$$

The rate of the reaction can be followed using a colorimeter. The methanoic acid is present in a large excess.

(a) Why can colorimetry be used to investigate the rate of this reaction? (1 mark)

e Colorimetry is used to investigate the rate of a reaction if a reactant or product is coloured.

> **Student answer**
>
> **(a)** The orange colour of bromine fades ✓ as the reaction proceeds and colourless bromide ions are produced.

e **1/1 mark awarded** Look carefully at the reaction equation and determine if there is a coloured substance present — an aqueous solution of bromine is orange. As the reaction proceeds the colour of the bromine fades and can be monitored by colorimetry.

(b) Describe how colorimetry could be used to determine the order of reaction with respect to bromine. (4 marks)

> **(b)** Measure absorbance against time ✓.
>
> Use a calibration curve of known concentrations of bromine to convert absorbance to concentration ✓.
>
> Plot bromine concentration against time ✓ and determine order from shape/find half-lives ✓.
>
> *Or*
>
> Find gradients of tangents at points ✓ and plot rate against bromine concentration and determine order from shape of graph ✓.

e **4/4 marks awarded** From experimental data there are different ways of determining the order: plotting $[Br_2]$ against time and using the shape of the graph to determine the order with respect to Br_2 (and verifying first order using half-life calculations); or take tangents to the $[Br_2]$ against time graph at various concentrations and plot a rate vs $[Br_2]$ graph to identify the order.

(c) Suggest an alternative method for investigating the rate of this reaction. (1 mark)

ⓔ Again look carefully at the reaction equation — carbon dioxide gas is also produced, so measure the gas volume against time.

(c) Use a gas syringe to measure the volume of gas produced against time ✓.

(d) Suggest a suitable chemical to use as the catalyst for the reaction. (1 mark)

(d) Any strong acid such as hydrochloric ✓.

(e) Explain the purpose of adding a large excess of methanoic acid. (2 marks)

(e) The concentration of methanoic acid is constant ✓. The effect of changing the concentration of bromine on the rate can be determined ✓.

ⓔ **2/2 marks awarded** Adding a large excess of methanoic acid means that its concentration stays constant during the reaction, allowing the effect of changing the concentration of the bromine concentration to be determined.

(f) Use the results in the table below to plot a graph of bromine concentration against time. (2 marks)

Time/s	$[Br_2]/10^{-3}\,mol\,dm^{-3}$	Time/s	$[Br_2]/10^{-3}\,mol\,dm^{-3}$
0	10.0	180	5.3
10	9.0	240	4.4
30	8.1	360	2.8
90	6.7	480	2.0
120	6.2	600	1.3

(f)

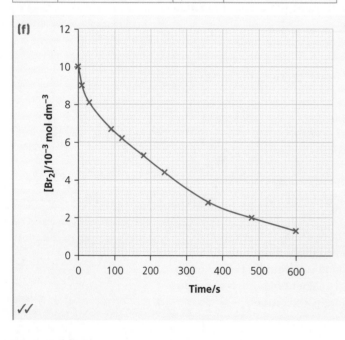

✓✓

(g) Determine two values for half-lives for the reaction from your concentration–time graph and determine the order of the reaction. (2 marks)

ⓔ From the concentration–time graph the shape is similar to that of a first-order reaction. To find the half-life, determine the time taken for the concentration to fall from 1.3 to 0.65 etc.

(g) All half-lives are around 200 ✓; first order ✓.

Question 19

The equations for reactions A and B are:

■ Reaction A: $2H_2O_2(aq) \rightarrow 2H_2O(l) + O_2(g)$

■ Reaction B: $CH_3COCH_3(aq) + I_2(aq) \rightarrow CH_3COCH_2I(aq) + H^+(aq) + I^-(aq)$

Choose which of the following apparatus could be used to continuously monitor the rate of (a) reaction A and (b) reaction B. (2 marks)

balance   colorimeter   gas syringe   pH meter

Student answer

(a) Reaction A: balance and gas syringe ✓

(b) Reaction B: colorimeter and pH meter ✓

ⓔ **2/2 marks awarded** In reaction A a gas is produced so a gas syringe or a balance could be used. The hydrogen peroxide and water are both colourless, so a colorimeter could not be used.

In reaction B iodine is coloured, so a colorimeter could be used. Hydrogen ions are generated, so a pH meter could be used.

Question 20

An inorganic compound contains a cation that is either NH_4^+, Fe^{3+} or Cu^{2+} and an anion that is Cl^-, SO_4^{2-} or CO_3^{2-}.

Plan a series of tests that you could carry out on samples to identify the ionic compound. Your tests should produce at least one positive result for each ion. For each test:

■ include details of reagents used, relevant observations and equations

■ explain how your observations allow the ions to be identified

You may include flow charts or tables in your answer. (8 marks)

ⓔ The anion tests should be carried out in the order below. Otherwise dilute nitric acid can be added before the silver nitrate and barium nitrate. Make sure you read the question carefully. You need to include reagents, observations and equations. It is best to give ionic equations for ion tests.

For the cation tests:

- Add aqueous sodium hydroxide ✓.
- Fe^{3+} brown ppt:

  $Fe^{3+} + 2OH^- \rightarrow Fe(OH)_3$ ✓
- Cu^{2+} blue ppt:

  $Cu^{2+} + 2OH^- \rightarrow Cu(OH)_2$ ✓
- Warm a sample with sodium hydroxide solution and test the gas released with moist red litmus and it should turn blue, indicating that alkaline ammonia has been produced:

  $NH_4^+ + OH^- \rightarrow NH_3 + H_2O$ ✓

For the anion tests:

- This should be carried out in order ✓.
- Add nitric acid. If there is effervescence it is a carbonate:

  $CO_3^{2-} + H^+ \rightarrow H_2O + CO_2$ ✓
- Add barium nitrate. If there is a white ppt then it is a sulfate:

  $Ba^{2+} + SO_4^{2-} \rightarrow BaSO_4$ ✓
- Add silver nitrate. If there is a white ppt it is a chloride:

  $Ag^+ + Cl^- \rightarrow AgCl$

  Add some dilute aqueous ammonia and the ppt should dissolve ✓.

Question 21

A pH meter was used to investigate the effect of dilution on the dissociation of ethanoic acid. The solutions shown in the table were prepared by diluting a $0.10 \, \text{mol dm}^{-3}$ solution of the acid. A pH meter was calibrated and the pH of each solution measured, starting with the least concentrated.

Concentration of ethanoic acid/mol dm^{-3}	pH reading	Calculated pH for solutions of hydrochloric acid of the same concentration
0.00010	4.2	
0.00100	3.5	
0.01000	3.0	
0.10000	2.7	1.0

(a) Describe how a $0.010 \, \text{mol dm}^{-3}$ solution of ethanoic acid was prepared from the $0.10 \, \text{mol dm}^{-3}$ solution.

(3 marks)

(a) Rinse a pipette with the $0.010 \, \text{mol dm}^{-3}$ ethanoic acid and then use a safety pipette filler with the pipette to transfer $25.0 \, \text{cm}^3$ of the solution ✓ into a $250 \, \text{cm}^3$ volumetric flask ✓. Make up with deionised water until the bottom of the meniscus is on the zero mark at eye level ✓.

ⓔ **3/3 marks awarded** Dilutions are made up in a volumetric flask. The dilution factor is ×10, so 25.0 cm³ is diluted to 250.0 cm³.

(b) The water used for the dilutions had been boiled and then allowed to cool to room temperature. Explain why this improved the accuracy of the measurements. (2 marks)

(b) Boiling removes dissolved carbon dioxide ✓, which is acidic and may affect the pH of the water ✓.

ⓔ **2/2 marks awarded** Deionised water contains acidic carbon dioxide. This may affect the pH of water.

(c) Explain why it was necessary to calibrate the pH meter. (1 mark)

(c) After storage, a pH meter does not give accurate readings because the glass electrode in the pH meter does not give a reproducible EMF over longer periods of time ✓.

(d) Why did the students measure the pH of the most dilute solution first and then work up to the more concentrated solutions in order? (1 mark)

(d) The pH meter responds to the presence of hydrogen ions. If the most concentrated solution was measured first, it would be difficult to ensure that all of the hydrogen ions had been rinsed away with deionised water, and this could affect the result ✓.

(e) Describe how the pH was measured. (2 marks)

(e) Place the pH probe in the solution, record the value to 1 decimal place ✓. Remove the probe and wash with deionised water ✓.

(f) Calculate the pH values for hydrochloric acid which are missing from the table. (1 mark)

ⓔ Hydrochloric acid is a strong acid so use the equation $pH = -log[H^+]$

(f) 4.0, 3.0, 2.0 ✓

(g) What do the differences in pH between each acid at each concentration show about the effect of dilution on the degree of dissociation of ethanoic acid? (1 mark)

(g) The values get closer as the acids get more dilute/the degree of dissociation increases with dilution ✓.

ⓔ **1/1 mark awarded** The pH values of ethanoic acid get closer to the pH values of hydrochloric acid as the acids get more dilute, showing that the degree of dissociation of the weak acid increases as it is diluted.

Question 22

The melting point of paracetamol is 169°C. Four students, A–D, each made paracetamol in the laboratory and recorded the melting point. Which student(s) made the purest paracetamol? Explain your answer.

(1 mark)

Student	A	B	C	D
Melting point /°C	160–166	160–164	167–168	169–171

ⓔ The purest solid will have the narrowest range, though if it is prepared in the lab it may still be lower than the true melting point.

Student answer

The answer is C as it has the narrowest range ✓.

Question 23

Methyl 3-nitrobenzoate is prepared by the following method:

- Weigh out 2.7 g of methyl benzoate in a small conical flask and then dissolve in 5 cm³ of concentrated sulfuric acid.
- When the solid has dissolved cool the mixture in ice.
- Prepare the nitrating mixture by carefully adding 2 cm³ of concentrated sulfuric acid to 2 cm³ of concentrated nitric acid and then cool this mixture in ice.
- Add the nitrating mixture drop by drop to the solution of the methylbenzoate, stirring with a thermometer and keeping the temperature below 10°C.
- When the addition is complete allow the mixture to stand at room temperature for 15 minutes.
- Pour the reaction mixture onto 25 g of crushed ice and stir until all the ice has melted and crystalline methyl 3-nitrobenzoate is formed.
- Filter the crystals using Buchner filtration, wash with cold water, recrystallise from ethanol and obtain the melting point.

(a) What homologous series does methyl benzoate belong to? What is the molecular and the empirical formula of methyl benzoate?

(3 marks)

Student answer

(a) Ester ✓; $C_6H_5COOCH_3$ ✓; C_4H_4O ✓

ⓔ 3/3 marks awarded Methyl benzoate is an ester because it contains an ester bond, –COOR. Remember that benzene is C_6H_6 but the H is substituted for $COOCH_3$, so the molecular formula is $C_6H_5COOCH_3$. To find the empirical formula, collect together each element — $C_8H_8O_2$ — and find the simplest ratio; in this case divide by 2.

(b) What conditions were used during this preparation to prevent further nitration of the product to the dinitro derivative? (1 mark)

(b) The temperature was kept low (less than 10°C) to prevent further nitration ✓.

(c) How do the conditions for the nitration of methyl benzoate differ from those for the nitration of benzene? (1 mark)

(c) Benzene is heated to 50°C, while methyl benzoate is cooled to less than 10°C ✓.

ⓔ 1/1 mark awarded You should know that benzene is nitrated by heating under reflux at 50°C .

(d) How do you expect the melting point of the impure sample to compare with the recrystallised sample? (2 marks)

(d) Lower ✓; wider range ✓

ⓔ 2/2 marks awarded Recrystallisation purifies a solid, so the impure sample should have a lower melting point and a wider range.

(e) Suggest ways in which the product may be lost during the experimental procedure. (2 marks)

(e) Some product is lost in purification steps ✓. Some product is lost in transferring between vessels ✓. Production of side products ✓. Not all starting material reacted ✓. *(Any 2)*

(f) Why were the crystals washed with cold water? (1 mark)

(f) To remove aqueous impurities, such as acid ✓.

ⓔ 1/1 mark awarded Cold water will dissolve aqueous impurities, such as the concentrated acids.

Question 24

Isopropyl acetate is an ester with boiling point of 88°C. It can be prepared by heating acetic acid with isopropyl alcohol in the presence of a catalyst in a round-bottomed flask. The product is removed from the flask, purified and analysed.

Isopropyl alcohol

(a) (i) Write an equation for the equilibrium reaction between acetic acid (ethanoic acid) and isopropyl alcohol. (1 mark)

ⓔ The product is an ester, which has functional group –COO, made from ethanoic acid and the isopropyl alcohol. In this reaction water is removed.

(ii) Name the catalyst added to the flask. (1 mark)

(iii) What is the name of the method of heating used, and what else should be added to the round-bottomed flask? (2 marks)

Student answer

(a) (i) $CH_3COOH + (CH_3)_2CHOH \rightleftharpoons CH_3COOCH(CH_3)_2 + H_2O$ ✓

(ii) Concentrated sulfuric acid ✓

ⓔ **1/1 mark awarded** In an esterification reaction the catalyst is usually a concentrated mineral acid, such as sulfuric acid.

(iii) Reflux ✓; anti-bumping granules ✓

ⓔ **2/2 marks awarded** Organic liquids should be heated under reflux so that there is no loss of product, and anti-bumping granules should be added to the flask to ensure even boiling.

(b) Give the IUPAC name for isopropyl alcohol. (1 mark)

ⓔ There are three carbons in the longest chain so prop- is used and the –OH group is on the second carbon.

(b) Propan-2-ol ✓

(c) Assuming a 40% yield, what is the minimum mass of isopropyl alcohol required to produce 10.2 g of isopropyl acetate? (2 marks)

ⓔ First calculate the moles of isopropyl acetate and then the mass of the alcohol. It is only a 40% yield, so to ensure 10.2 g of product multiply the mass of alcohol by 100/40.

(c) moles of ester = 10.2/102 = 0.1

moles of alcohol = 0.1

mass of alcohol = 0.1 × 60 = 6 g

6 × 100/40 = 15 g ✓✓

(d) Name the experimental technique used to remove the product from the flask. (1 mark)

(d) Distillation ✓

🟢 **1/1 mark awarded** The product will be mixed in the reaction flask with the catalyst and any unreacted reactants. All are liquids, so it is separated by distillation.

(e) Name a solution that can be added to remove acidic impurities from the crude product. (1 mark)

(e) Sodium carbonate solution ✓

🟢 **1/1 mark awarded** An aqueous solution of sodium carbonate will remove any acidic impurities, and it must be added in a separating funnel.

(f) Giving experimental details, describe how the crude product can be purified using anhydrous calcium chloride. (3 marks)

(f) Swirl with anhydrous calcium chloride ✓ until the liquid goes clear ✓. Filter off the solid drying agent/decant off the crude product ✓.

🟢 **3/3 marks awarded** Anhydrous calcium chloride is a solid. It is a drying agent and can be added to the crude product in a flask, until the product changes from cloudy to clear, and it is then filtered off.

(g) Suggest how infrared spectroscopy could be used to show that the product did not contain any unreacted acetic acid or isopropyl alcohol. (1 mark)

(g) There would be no –OH absorption ✓.

🟢 **1/1 mark awarded** Both acetic acid and the alcohol contain an –OH group. If they are not present, there should be no –OH absorption.

(h) State the integration pattern in the NMR spectrum of isopropyl acetate. (1 mark)

🟢 There are three types of chemically equivalent hydrogen atom. The integration gives the ratio of each type of chemically equivalent hydrogen.

(h) The integration pattern is $3:1:6$ ✓.

Question 25

Methanal is gaseous at room temperature but can be obtained as a solution that contains 37.0% methanal, by mass. Methanal causes severe irritation of the nasal system. It can be reacted with excess aqueous 2,4-dinitrophenylhydrazine to form a hydrazone, as shown in the reaction below.

(a) Calculate the mass of methanal solution needed to form 1.4 g of the 2,4-dinitrophenylhydrazone, assuming a 95% yield.

(4 marks)

ⓔ First find the relative molecular masses — M_r CH_2O = 12 + 2 + 16 = 30 and M_r $C_7H_6N_4O_4$ = 210. Then find the moles of hydrazone, moles of methanal by ratio and the mass of methanal. The methanal solution contains 37% methanal by mass.

> **Student answer**
>
> **(a)** moles = $\dfrac{1.4}{210}$ = 0.006 6667 ✓
>
> mass of methanal = 0.006 6667 × 30 = 0.2000 g ✓
>
> The methanal solution contains 37.0% methanal by mass, so the mass needed is:
>
> $\left(\dfrac{100}{37}\right)$ × 0.2 = 0.54 g ✓
>
> For a 95% yield:
>
> 0.54 × $\left(\dfrac{100}{95}\right)$ = 0.57 g ✓

(b) Explain the safety precaution that needs to be followed during the preparation.

(1 mark)

> **(b)** Use a fume cupboard as methanal irritates the nose ✓.

(c) The 2,4-dinitrophenylhydrazone is formed as an orange precipitate, which is collected by suction filtration using a Buchner flask.
  (i) Explain how 'Buchner filtration' is carried out. (3 marks)
  (ii) State why it is used in preference to normal filtration. (2 marks)

> **(c) (i)** Place filter paper into a Buchner funnel ✓.
>   Place the funnel into a Buchner flask ✓.
>   Suck the air through/attach to pump ✓.
>
> **(ii)** It is faster ✓. It produces drier crystals ✓.

(d) The solvent used to recrystallise the 2,4-dinitrophenylhydrazone is ethanol. Explain why the 2,4-dinitrophenylhydrazone is soluble in ethanol and not in hexane.

(2 marks)

> **(d)** Hydrazone can form hydrogen bonds with ethanol ✓. Hexane is non-polar and cannot form hydrogen bonds with hydrazone ✓.

(e) Describe, giving experimental details, how the 2,4-dinitrophenylhydrazone is recrystallised.

(4 marks)

ⓔ Note that the stem of the question states that the solvent is ethanol, so make sure you mention ethanol in your answer.

(e) Dissolve in the minimum volume of hot ethanol ✓. Filter when hot ✓. Allow to cool and crystallise ✓. Filter under reduced pressure ✓.

Question 26

Which one of the following techniques could be used to accurately determine the percentage iron(II) ion content of an iron sulfate tablet?   (1 mark)

A   thin-layer chromatography

B   melting-point determination

C   addition of a neutral solution of iron(III) chloride

D   titration with potassium manganate(VII)

Student answer

The answer is D ✓.

🔵 **1/1 mark awarded** Thin-layer chromatography does not determine quantitatively the iron(II) content. Melting-point determination shows purity, not the percentage content.

Knowledge check answers

1 Down the sink is safe because the product is a neutral solution of sodium chloride.

2 $7.9\,g\,cm^{-3}$

3 It absorbs water from the atmosphere.

4 mass lost = $1.24\,g$
overall uncertainty = $2 \times 0.005\,g$
percentage uncertainty in mass loss
$= \left|\dfrac{(2 \times 0.005)}{1.24}\right| \times 100 = 0.81\%$

5 pink to colourless

6 $CH_4 + 2O_2 \rightarrow CO_2 + 2H_2O$

7 mass of spirit burner + methanol; mass of spirit burner + methanol at end; mass of water

8 To prevent draughts and so prevent heat loss.

9 carbon monoxide; carbon (soot)

10 Only hydrogen and hydroxide ions are involved in the reaction — $H^+ + OH^- \rightarrow H_2O$ — so the nature of the strong acid and alkali does not matter.

11 More particles are present in the same volume, and so there are more collisions with the activation energy per second.

12 Warm the sample with sodium hydroxide solution. A pungent-smelling gas is released, which changes moist indicator paper to blue and gives white fumes with a glass rod dipped in conc. hydrochloric acid.

13 white ppt

14 white ppt; very thin white ppt

15

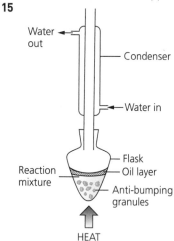

Water out

Condenser

Water in

Flask

Reaction mixture

Oil layer

Anti-bumping granules

HEAT

16 anhydrous sodium sulfate/anhydrous magnesium sulfate/anhydrous calcium chloride

17 Ethanoic acid is soluble in water.

18 reflux

19 The bromine adds on to the carbon–carbon double bond.

20 silver ion; $Ag^+ + e \rightarrow Ag$

21 $2CH_3CH_2COOH + Na_2CO_3 \rightarrow 2CH_3CH_2COONa + H_2O + CO_2$
$2CH_3CH_2COOH + Mg \rightarrow (CH_3CH_2COO)_2Mg + H_2$
effervescence

22 percentage error = $\left|\dfrac{0.01}{0.12}\right| \times 100 = 8.3\%$

percentage error = $\left|\dfrac{1}{120}\right| \times 100 = 0.8\%$

23 $50.0\,cm^3$ burette

24 colorimeter

25 The time taken for the concentration of one of the reactants to fall by half.

26 C — first order; D — second order; $mol^{-2}\,dm^6\,s^{-1}$

27 s^{-1}

28 rate = $k[H_2][NO]^2$

29 A temperature of $298\,K$; gases at a pressure of $100\,kPa$; solutions at a concentration of $1.0\,mol\,dm^{-3}$.

30 a $Mg + CuSO_4 \rightarrow MgSO_4 + Cu$
$Mg \rightarrow Mg^{2+} + 2e^-$ Mg is oxidised
$Cu^{2+} + 2e^- \rightarrow Cu$ Cu^{2+} is reduced

b $Cl_2 + 2KBr \rightarrow Br_2 + 2KCl$
$Cl_2 \rightarrow 2Cl^- + 2e^-$ Cl_2 is oxidised
$2Br^- \rightarrow Br_2 + 2e^-$ Br^- is reduced

31 A strong acid is one that dissociates into ions completely, in solution. A weak acid is one that partially dissociates in solution. A buffer is a substance that resists changes in pH when small amounts of acid or alkali are added, or when diluted.

32 recrystallisation

33 It would be lower than that of the pure substance and with a wider range.

34 percentage yield = $\left|\dfrac{\text{actual yield}}{\text{theoretical yield}}\right| \times 100$

35 green ppt with iron(II) ions; brown ppt and bubbles of gas with iron(III) ions

Index